AUSTRALIA'S GOLD RUSHES

AUSTRALIA'S GOLD RUSHES

ROBERT COUPE

Published in Australia by
New Holland Publishers (Australia) Pty Ltd
Sydney • Auckland • London
Unit 1/66 Gibbes St Chatswood NSW 2067
5/39 Woodside Ave Northcote Auckland 0627 New Zealand
131-151 Great Titchfield Street London WIW 5BB United Kingdom

First published in 2000
Reprinted in 2006, 2012, 2018

National Library of Australia Cataloguing-in-Publication Data:

Coupe, Robert.
Australia's gold rushes.

Includes index
ISBN 9781864365474

1. Gold mines and mining – Australia – History.
2. Australia – gold discoveries.
3. Australia – History – 1851–1891.
I. Title

994.031

Cartography: DiZign Pty Ltd
Reproduction: Dot'n'Line
Printer: Toppan leefung Printing Limited

Cover photos
Front cover: Alluvial gold washing.
Back cover: The gold mining town, Gulgong, in New South Wales, around 1872.

Picture Acknowledgements
All images © National Library of Australia with the exception of:
— Image Library, State Library of NSW: p 11
— La Trobe Collection, State Library of Victoria: pp 16, 19, 24, 25, 33, 37, 39
— Coo-ee Historical Picture Library: front cover, pp 31, 54 (bottom)
— Photo Library of Australia: p 6.

Contents

In Quest of Gold 6
Early Finds 8
Off to California! 10
The Start of a Gold Rush 12
Spreading Fever 14
Gold Further South 16
Ballarat 18
Mount Alexander 20
Bendigo 22
Further Afield 24
New South Wales Again 28
Northern Goldfields 32
New Arrivals 38
Troubles on the Goldfields 42
Anti-Chinese Feelings 47
The Eureka Stockade 51
The Golden West 56
Aspects of Goldfields Life 61
Index 64

IN QUEST OF GOLD

Throughout the centuries people in most civilisations have quested for gold. They have dreamed of possessing it, fought for it, travelled vast distances in search of it, and risked, and often lost, their lives in attempts to find it. Powerful nations have even conquered other peoples in order to get hold of it, and for hundreds of years a legend about gold sent

Tutankhamen's golden mask.

expedition after expedition in search of a non-existent hoard of golden riches.

Ancient treasures

Almost 5000 years ago the ancient Egyptians mined gold in the desert sands. They made jewellery with it, they lined the coffins of their dead pharaohs and other wealthy people with thin sheets of it, they buried golden ornaments in tombs and pyramids to comfort the dead in their afterlife, and they fashioned statues of their gods with it. It is even possible that the tops of some of their pyramids were covered with a gleaming cap of gold. When, in 1922, the British archaeologist Howard Carter discovered the tomb of the pharaoh Tutankhamen, who had died more than 3000 years earlier, he was amazed by the sight of gold. 'Everywhere', he later wrote, he could see 'the glint of gold'. One of the coffins in the royal tomb was made of solid gold and the pharaoh's mummy was covered with a magnificent mask of gold.

About 50 years before Carter's discovery, the German archaeologist Heinrich Schliemann unearthed a 3000-year-old hoard of 10 000 golden objects at the ancient city of Troy in Turkey. He also found many golden treasures in tombs at Mycenae in southern Greece. One of these was a gold mask that Schliemann (wrongly as scholars later showed) thought showed the face of the Greek king Agamemnon. If you visit Greece you can see this wonderful object in a museum in Athens.

The legend of El Dorado

In English, an 'Eldorado' is a place of almost unimaginable riches. In Spanish the phrase means 'the golden man'. When Spanish invaders conquered the Aztec people of Mexico and the Incas of Peru in the sixteenth century, they found civilisations that were rich in gold. Large Spanish ships, called galleons, crossed the oceans to bring these newly found treasures back to Spain. In 1537 a Spanish band of invaders conquered the Chibcha peoples who lived near the present city of Bogota in Colombia. Here they found fields of cultivated crops and villages where houses were decorated with sheets of gold. One of the Chibcha people told the newcomers about a lake in the nearby mountains, where the local chief threw gold and precious stones as an offering to the gods. As well, every year the chief would have his body covered with fine gold dust and would dive into the water to add this gold to the treasures at the bottom of the lake.

The Spaniards believed the man and stories began to spread of a mountain lake with treasure on its bed and even of a city with streets and buildings of gold. Fortunes were spent in the search for El Dorado. Although pieces of gold were found around the edges of a lake called Lake Guatavita, nothing like the fabled treasure was found. Eventually, at the beginning of the twentieth century, almost 400 years after the Spanish invasions, a British expedition pumped Lake Guatavita almost dry. They did find some gold objects, but not nearly enough to cover the costs of the search.

For several decades from the 1850s, different parts of Australia became an Eldorado for thousands of gold seekers, many of whom came from distant parts of the world in the hope of making their fortune. Many did; many others found that fortune was against them.

Rich and rare

Why are people lured by gold? Part of the answer lies in the sheer beauty of the metal. Part of it, too, is the tradition, established over thousands of years, that associates gold with wealth and power. Perhaps most important, though, is that it is rare. People have worked out that in the 5000 years or so since humans started mining it, only about 120 000 tonnes of the precious metal have been found. If all the gold that has ever been produced were put into one solid mass, it would take up only about 7000 cubic metres of space. That's about as big as a small block of flats.

It's really no wonder, then, that at various times in history, large numbers of people have caught 'gold fever'. One of those times began near the town of Bathurst in New South Wales just about 150 years ago.

EARLY FINDS

Edward Hargraves is often given the credit for being the first person to discover gold in New South Wales. In January 1851 this Englishman, who had just returned from looking for gold in California, rode from Sydney over the Blue Mountains to Bathurst, expecting to find gold in the area. At the town of Guyong, north-west of Bathurst, he met up with John Lister, the son of the publican there. Together they went panning for gold and at a place called Lewis Ponds Creek they managed to unearth a total of 16 specks of gold. It wasn't much, and was certainly worth almost nothing, but it was enough to get Hargraves excited. Hargraves had a strong desire to get rich quickly, but was not keen to spend the time and effort needed to find larger amounts of gold. He hurried back to Sydney, determined to make the most of his meagre find.

He managed to arrange a meeting with the colonial secretary for the colony of New South Wales, Deas Thomson, and proudly displayed his gold. He claimed he had discovered a great goldfield and asked for a reward of at least £500. He was granted a sum of £30 to help cover expenses and asked to take a government surveyor to his goldfield. On 7 April, while Hargraves was still in Sydney, John Lister and two neighbours, James and William Tom, found two nuggets of gold in the same creek. They were the first to find 'payable' gold—that is, gold that was worth selling. But when they wrote to Hargraves about their exciting discovery, their cunning friend made certain, as we shall learn later, that it was he who got the credit, and the reward. We will read about that on pages 12 and 13.

Gold on the ground

It is possible that hundreds of people in New South Wales had picked up gold before Hargraves and Lister made their discovery. As settlement spread outwards from Sydney, shepherds wandered the countryside grazing and guarding their employers' sheep. It is almost certain that some of these found pieces of gold. Many of them would not have recognised the gold they found for what it was, especially if it was embedded in quartz rock. Others might have decided to say nothing about it. After all, they may not have been allowed to keep it, because there was a law that gold and other minerals belonged to the government, no matter who found them.Some might have quietly sold their gold for a little extra money.

There was no gold rush before 1851 largely because very few people would have had the time to go in search of gold. Almost no-one would have known where to start looking. The governors and other authorities, too, had their reasons for keeping quiet about any finds of gold. In a society with large numbers of convicts and ex-convicts they felt that law and order could break down very quickly if gold fever gripped the population.

The first find of gold that we know about was made not far from Bathurst, in 1823, when a surveyor, James McBrien, noted in his official book that he had spotted some pieces of gold. There was so little of it that McBrien considered it of no importance. The Polish explorer, Paul Edmund de Strzelecki, who was a keen amateur geologist, also found some specks of gold—in rocks near Wellington, north of Bathurst. He gathered some rock samples and took them to England. The well-known geologist, Sir William Murchison, examined them and suggested that there could be payable gold in the region from which they were collected. However, his views were not made widely known in the colony.

A few years later, in 1847, William Clarke, an Anglican clergyman and teacher, who was also a trained geologist, found some tiny pieces of gold in the Blue Mountains. It was not much of a find, and in any case Clarke did not want to make much of it. He did write to the *Sydney Morning Herald* about his find, but made the point that he did not believe it was payable gold. He didn't want it to be, because he

Hargraves and Lister at Lewis Ponds Creek.

strongly believed that greed was evil and that a gold rush would create widespread greed. Clarke showed his gold to the governor, George Gipps, who, it seems, was not very interested in it. The governor certainly had no desire to start a rush in search of gold.

OFF TO CALIFORNIA!

New South Wales was founded as a colony of Great Britain in 1788, when the 11 ships of the First Fleet arrived with their cargo of convicts and soldiers. Convicts continued to arrive in Sydney and other parts of New South Wales for another 50 years. But by 1830 Sydney was no longer just a convict town. Many men and women who had completed their sentences stayed on and prospered as free citizens; many others had arrived from Britain of their own accord to make their fortune in a new country. Others had taken flocks of sheep far into the country and established themselves as squatters. New South Wales was quite a prosperous place in 1830—so much so that advertisements in English newspapers encouraged tradespeople to come and settle in the young colony, promising them that work was plentiful and wages good. This prosperity was probably part of the reason that there was no widespread interest in finding gold in the colony.

A change of fortune

Early in the 1840s, however, the colony suffered a reversal of fortune. Wool was the main source of wealth, and in 1842 the price of wool fell dramatically as demand for it in British woollen mills fell away. To make matters worse the colony was in the grip of a fierce drought. Many shepherds and other rural workers lost their jobs, many people on the land were ruined and city businesses were forced to close down. Although the situation slowly improved later in the decade, times were still hard as Christmas approached in 1848. When, in December 1848, the first reports of gold discoveries and fortunes being made 12 000 kilometres away across the Pacific Ocean reached Sydney, many people's interest was aroused.

In 1848 few people in New South Wales had even heard of California. This area on the west coast of North America had only just been won by the United States after a war with Mexico. In January 1848 an employee of an American sawmill owner discovered gold while digging a water course in the Sierra Nevada mountains, east of San Francisco. News of the find spread widely and in a few months, thousands of fortune-seekers from other parts of the United States and Europe were making for California.

People in the colony were at first slow to get off the mark. By the end of January 1849 fewer than 100 people had sailed from Sydney, even though enterprising ship owners had been quick to advertise cheap fares to San Francisco. At first many did not trust the reports, some of which were wildly exaggerated. As well, news in those days took some time to get through to country districts, where many of the more adventurous people were. By the end of June almost 400 New South Wales people had set sail, as well as others from South Australia and Van Diemen's Land, as Tasmania was then known. The move to California was now well under way.

Hargraves in California

On 20 July 1849 the sailing vessel *Elizabeth Archer* sailed out of Sydney Harbour bound for San Francisco, after being held up for three days by wild winter weather. Among the 150 or so passengers on board was Edward Hargraves, who was down on his luck and very keen to improve it. This tall, bulky 34-year-old had come to New South Wales 15 years earlier and had married a well-to-do merchant's daughter. His attempts to make a success of farming had failed, but he and his wife ran an inn north of Sydney and owned some cattle. Hargraves sold the cattle to pay for his passage in a comfortable cabin and left his wife and several children to go to California.

By all accounts, Hargraves neither worked very hard nor found much gold on the various Californian goldfields that he visited, although he later claimed to have struck it rich. But he looked and listened. He learned how to prospect for gold and noted the kind of country in which it might be found. He knew the country around Bathurst and made up his mind that gold was there to be found. He even wrote about it to a friend in Sydney, but was careful not to say exactly where he had in mind. Early in 1851 he sailed back to New South Wales, this time determined to find fame and fortune.

The Joseph Conrad *was typical of the ships which carried gold diggers to California.*

THE START OF A GOLD RUSH

Excited reports of fortunes made in California meant that gold was a frequent topic of conversation in the Australian colonies. Hargraves knew this and used it to his advantage. When, in Sydney, he received the letter from John Lister telling that they had discovered payable gold, he headed straight to the Listers' inn at Guyong to see for himself. Convinced now that there was a fortune to be won, the wily Hargraves made his plans.

He probably knew that he did not have the energy or the patience to prospect seriously for gold. He also understood that even if he had, he may not find much. What he wanted was to receive a reward from the authorities and to become famous as the discoverer of gold. It was in his interests, then, to start a rush for gold. Instead of keeping quiet about the discoveries, as Lister and the Tom brothers wanted, he started talking about it around the region. He gave a public lecture in Bathurst, telling his audience where he and his companions had found the gold, and giving advice on how to pan for it. He even led an expedition from Bathurst to 'Ophir', as he rather grandly named the area—Ophir in the Bible was the site of King Solomon's gold mines.

The arrivals begin

News spread quickly in the Bathurst region. At first a few men turned up to dig and pan for gold. Hargraves issued them with pieces of paper giving the right to prospect. These were the first gold licences in the colonies, but they were, of course, completely illegal. As reports of good finds got out, it was not long before several hundred fortune-seekers had found their way to the diggings at Ophir.

News was slower to reach Sydney. It was almost 300 kilometres away and roads in those days were difficult and dangerous. When it did arrive the authorities were not concerned at first, believing the reports were exaggerated. But when, a few days later, *The Sydney Morning Herald* reported the situation, the governor, now Charles FitzRoy, became alarmed, fearing that trouble would break out. After all, reports from California not only told of fortunes made; they also described violent quarrels, shootings and stabbings. Officials were sent to Ophir to tell the diggers that they had no right to be there, and that any gold they found belonged not to them, but to the Crown. The diggers ignored them.

The authorities were at a loss for what to do. Sending soldiers or police might only provoke violence, and in any case they would probably be outnumbered by the diggers, many of whom were armed. In late May, the governor accepted advice from Hargraves. He would issue official licences, similar to the ones that had been used in California. The price was set at 30 shillings per month, and the licence entitled a miner to prospect for gold in a small area called a 'claim' and allowed him to keep for himself any gold he found on that claim.

To the surprise of many people, most miners accepted the licensing system and paid the fee, without too much protest, to the official who rode around the diggings accompanied by 10 armed guards. As a result, there was little violence at the Ophir diggings. But only three months after gold was first discovered at Ophir, the gold there was beginning to run out. Many of the new arrivals did not find the riches they were expecting and the governor feared that this could lead to trouble. He granted Hargraves a £500 reward for being the 'first discoverer of gold in Australia', and paid him a generous fee to go out and find new goldfields. Hargraves imagined that he would find new fields further south, around the Murrumbidgee River, but when he went there he found nothing.

But by now, gold had brought Hargraves the fame and fortune he wanted, even though he had never found a sizeable amount of it. Two years later the New South Wales government gave him the huge sum of £10 000 in

Early arrivals at the Ophir diggings.

recognition of his services, and the following year he was presented to Queen Victoria in London. Some years later, he was granted a yearly pension for life. But in 1890, the year before he died, Hargrave's reputation was damaged. A government inquiry found that he should not have been honoured as the first discoverer of payable gold. That honour, the inquiry stated, should have gone to John Lister and James and William Tom.

SPREADING FEVER

The government of New South Wales did not need Edward Hargraves to find new goldfields—the miners soon found them for themselves. Some who had successfully worked their claims at Ophir till there was no more gold to be found, and others who had come to Ophir and found little or nothing, and still others who wanted to avoid paying the 30 shillings for a licence, began to move further afield.

Only a few weeks after the gold rush at Ophir had begun, some miners were packing up their equipment and moving further to the north-east and trying their luck along the banks of the Turon River. News of good finds along a 30-kilometre stretch of this river soon arrived in Bathurst, and then in Sydney, and started a new rush of diggers to this part of the country. A township of tents grew up at Sofala, named after a rich goldfield in East Africa, and another at a place called Tambaroora, about 40 kilometres to the north-east of Sofala. In between, miners' tents were strung out along the river. Most of the gold here was close to the surface—it was 'alluvial' gold. Reasonable finds were rather easy to make, and many people in the area had sold their gold in Bathurst, even before government officials arrived in early July to demand licence fees and approve claims. By this time there were close to 2000 prospectors in the area. Although mining was uncomfortable in the cold and wet winter conditions, at least the winter rains had swelled the river and loosened the soil, making prospecting easier. Many of the people who poured into these recently discovered diggings had little more than simple spades or tin dishes with which to scrape the soil. Despite this, a lot of them managed to prosper.

Many diggers went to the goldfields with only what they could carry on their backs.

Only three months earlier, Hargraves, Lister and James Tom had moved through this region but had gone away empty-handed. At that time the weather had been hot and dry, and the creeks and rivers flowed only slowly. Running short of water, the three men had not put a lot of effort into their search.

The biggest find yet!

Today, Hargraves, named after Edward Hargraves, is a tiny village about 30 kilometres north of Tambaroora, which itself is now little more than a dot on the map. But for more than 20 years after July 1851, Hargraves was the centre of a fierce and furious gold rush. On 16 July the Bathurst Free Press excitedly reported that 'Bathurst is mad again! The delirium of golden fever has returned with increased intensity'. It then went on to describe how Jemmy Irving, an Aboriginal shepherd employed on a cattle run owned by a Dr William Kerr, near the banks of Meroo Creek, had struck an outcrop of quartz rock with a

New South Wales goldfields.

tomahawk and had discovered, embedded in the rock, a mass of glistening gold. He rushed excitedly back to the homestead and informed his employer. Not convinced at first, Kerr went back to the spot with Irving and was delighted to discover that his shepherd was not mistaken. The two men set to work and hacked out three large pieces of gold-bearing quartz. When, a couple of days later, this treasure arrived in Bathurst after a slow 50-kilometre journey along muddy roads on board a horse-drawn dray, the bank's scales showed that the gold alone weighed more than 35 kilograms, an amount worth more than £4000.

It was the largest single amount of gold yet unearthed. It not only sent Bathurst mad, it also excited Sydney when *The Sydney Morning Herald* reported it a few days later. And when, months later, the news broke in far-away London it set a few people thinking that there may be rich rewards to be had in a distant colony at the other end of the world.

Jemmy Irving, at least, had struck it rich, because Kerr rewarded him handsomely with flocks of sheep and cattle and land on which to run them.

GOLD FURTHER SOUTH

When gold-seekers in their thousands eventually did arrive from overseas, it was not to the Turon fields that most of them went. By that time even richer finds had been made further south, in the hills and mountains to the north-west of Melbourne.

When the first gold rushes to Ophir were happening in early 1851, Melbourne was still a town in New South Wales. But in July 1851, it became the capital city of the newly declared separate colony of Victoria. When it separated from New South Wales, Victoria had a population of almost 80 000 people. Melbourne was going ahead in leaps and bounds. There was a feeling of confidence in the air, new houses were being built at a great rate and shops and businesses were expanding. Although it was tiny compared to the city of today, one businessman described it in 1850 as an 'immense place'. Imagine, then, how concerned local merchants became when they saw their employees leaving in droves in the hope of striking it rich on the New South Wales goldfields. Country landholders were also alarmed that their shepherds and workers would leave their animals and farms unattended. The answer was to start a gold rush in the new colony.

Almost a gold rush

There might well have been a great gold rush in the area more than two years earlier if the man who was to become Victoria's first governor, Charles LaTrobe, had not put an end to it. At the end of January 1849, while the lure of gold in California was beginning to entice people away from the colony, a Melbourne newspaper announced that gold had been discovered in the Pyrenees Mountains, less than 200 kilometres north-west of Melbourne. A shepherd named Campbell had discovered a sizeable quantity on his employer's property and had brought it to a jeweller in Melbourne. The jeweller recognised its value and persuaded Campbell to show him where he had found it. Campbell agreed, but then changed his mind and ran away. But the news got out and within a week about 40 diggers had arrived in the area. They did not stay long. The authorities did not want a local gold rush at this stage, and LaTrobe, who was then the superintendent of the Melbourne area, sent in troops to get rid of them.

Governor LaTrobe.

Rich rewards

A group of Melbourne businessmen, headed by the mayor of Melbourne, decided to offer a reward of £200 for the discovery of a payable goldfield within 200 miles (320 kilometres) of Melbourne. The Victorian authorities, headed by Governor LaTrobe, also now saw the potential value of gold to their new colony. They soon increased this amount to the vast sum of £10 000. It was not long before people were claiming the rewards.

One of the earliest finds was at a place called Warrandyte, about 30 kilometres north-east of Melbourne, on the Yarra River. The finder was Louis Michel, who ran a hotel in Melbourne. Following Michel's find, there was a small rush to the area, but it did not last long. But Michel did receive a reward of £1000 from the government.

A bigger gold rush, which also soon ran out of steam, occurred at Clunes in the Pyrenees Mountains, not far from where the shepherd Campbell had made his find. The first find here was made by James Esmond, a timber-worker who had been to California. As the Clunes goldfield was growing in strength, Thomas Hiscock, the blacksmith in the town of Mount Buninyong, which was on the way to the diggings at Clunes, decided to hack into the local quartz rock and to prospect in a nearby creek. In early August 1851, Hiscock found gold in both locations and news of his find brought new prospectors into the Buninyong area. Both Esmond and Hiscock were rewarded for their efforts with payments of £1000 each.

Buninyong was on the edge of a pastoral property known as 'Ballarat'. Ballarat was an Aboriginal word that meant 'camping place'. Before long thousands of gold miners would be camped there.

A view of Melbourne in 1847.

BALLARAT

The first person to stumble on gold on Ballarat station was James Regan. Regan, together with a 75-year-old mate, John Dunlop, who had come to the colonies from England as a free settler, had trudged more than 50 kilometres from Geelong to the Buninyong diggings in early August 1851. After two weeks prospecting in the wet, wintry conditions, the men had found very little. Discouraged, they thought of going back home, but Dunlop, though physically weakened by the strain, still retained a strong sense of purpose. He persuaded Regan to walk almost another 50 kilometres to the diggings at Clunes, to see if things were better there.

Regan's journey took him across the Ballarat station and there, at a place that became known as Golden Point, he found a rich vein of gold. He rushed back to Buninyong and excitedly led his old friend to the place where he made his discovery. Others who heard about the find soon followed, and so began the biggest gold rush that the colonies had yet seen.

Keeping out convicts

The rush to the Ballarat diggings alarmed the Victorian authorities, just as the rush to Ophir had worried those in New South Wales. There was good reason for the concern. Melbourne was literally emptying out, as people, particularly men, abandoned the city for the goldfields. Business people and people on the land now saw the gold rush that they had wished for as a greater evil than the New South Wales one had threatened to be.

The rush to Ballarat, 1851.

Even worse, they were afraid that ex-nvicts, or even escaped convicts, from Van Diemen's Land, as Tasmania was then called, would soon invade Victoria, bringing crime to Melbourne and violence to the goldfields.

Almost as soon as the first gold was found, some people suggested that ex-convicts should be stopped from entering Victoria and in 1852 this became a law. It did not work, however, because by that time many Vandemonians, as early Tasmanians were called, had already arrived. In any case many ex-convicts managed to slip through even after the law was passed.

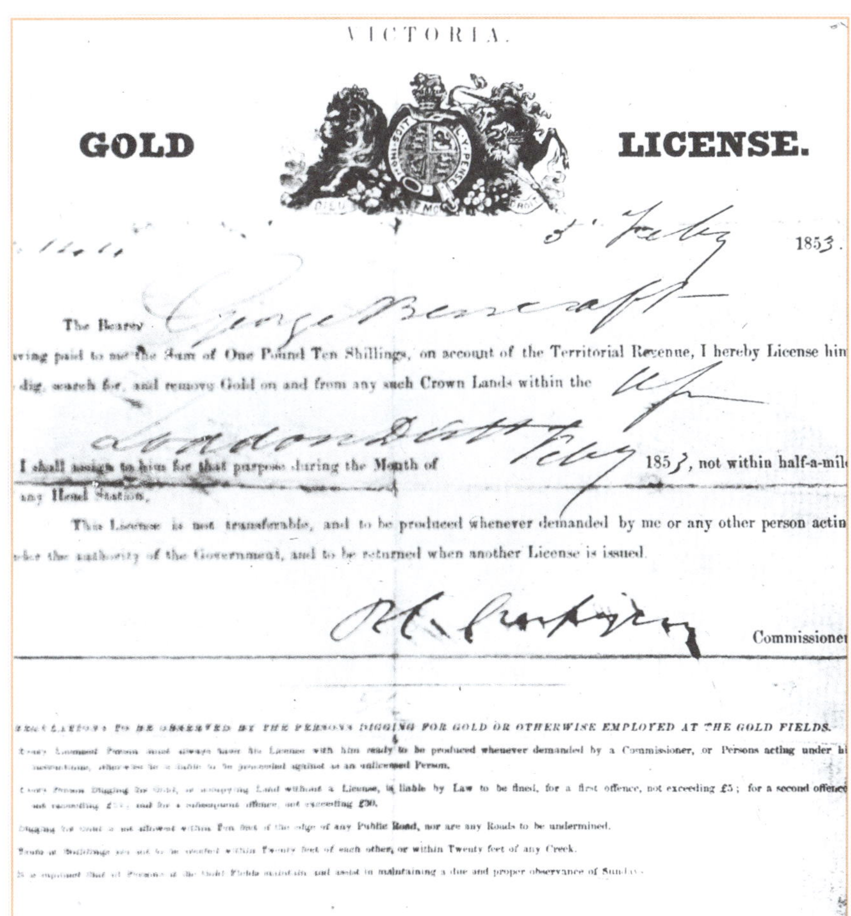

VICTORIA.

GOLD LICENSE.

[illegible] 5th July 1853.

The Bearer George Bencraft

having paid to me the Sum of One Pound Ten Shillings, on account of the Territorial Revenue, I hereby License him to dig, search for, and remove Gold on and from any such Crown Lands within the Up

London [illegible] July

I shall assign to him for that purpose during the Month of July 1853, not within half-a-mile of any Head Station.

This License is not transferable, and to be produced whenever demanded by me or any other person acting under the authority of the Government, and to be returned when another License is issued.

[illegible]

Commissioner

REGULATIONS TO BE OBSERVED BY THE PERSONS DIGGING FOR GOLD OR OTHERWISE EMPLOYED AT THE GOLD FIELDS.

Every Licensed Person must always have his License with him ready to be produced whenever demanded by a Commissioner, or Persons acting under his instructions, otherwise he is liable to be proceeded against as an unlicensed Person.

Every Person Digging for Gold, or occupying Land without a License, is liable by Law to be fined, for a first offence, not exceeding £5; for a second offence not exceeding £[illegible]; and for a subsequent offence, not exceeding £30.

Digging for Gold is not allowed within Ten feet of the edge of any Public Road, nor are any Roads to be undermined.

Tents or Buildings are not to be erected within Twenty feet of each other, or within Twenty feet of any Creek.

It is enjoined that all Persons at the Gold Fields maintain and assist in maintaining a due and proper observance of Sundays.

A Victorian gold licence.

Issuing licences

Less than a month after Regan had found his gold, there were about 1000 people camped on the Ballarat station. LaTrobe decided to control the diggings by issuing licences. Like those in New South Wales, these licences were for a month and were issued at a cost of 30 shillings. They entitled miners to work a square patch of ground with sides about 2.5 metres long. But when officials and soldiers arrived at Ballarat on 21 September 1851, they were not warmly welcomed. Many miners hurled abuse at them and tried to prevent others from buying licences. It was an ominous warning of much worse troubles that would later erupt. Governor LaTrobe, alarmed at the reports he received, hastily recruited more goldfields police and things at Ballarat soon settled down. It was just as well they did, because within a couple more weeks, there was a spreading tent city of more than 2000 people. By late October this number had risen to over 5000.

Hard work

Governor LaTrobe himself went to Ballarat soon after the first licences were issued and he actually saw several men become rich in a day. But he also saw what a gamble goldmining was. Most of the gold at Ballarat lay well beneath the surface, some of it tens of metres deep. While some prospectors found gold on the surface, panning along creeks with wooden cradles or tin dishes, others dug holes deep in the ground—a back-breaking job, and one that as often as not went unrewarded. In many cases trees had to be felled before digging could begin, and this only added to the frustration if no gold was found.

It is hardly surprising, then, that when news got out of another goldfield 65 kilometres to the north-east—one where, it was said, abundant gold was just below the surface—diggers left the Ballarat diggings almost as quickly as they had arrived.

Just three months after the Ballarat gold rush began, it had ended, and only two to three hundred hardy diggers remained at Ballarat. A year later, however, gold-seekers would be back in force.

MOUNT ALEXANDER

The new field was around Mount Alexander, a jutting rocky peak surrounded by heavily wooded rolling hills. The gold rush to the Mount Alexander region, which began in late October 1851, occurred a full three months after the first find had been made in the area. This is because the finders at first decided to keep the news to themselves.

In July 1851, Christopher Peters was a shepherd living on an out-station of a large sheep station. An out-station was a small settlement, consisting perhaps of one or two huts, on a remote part of a sheep station. Shepherds lived in these huts. They watched the sheep graze by day and herded them into pens at night. So they spent long hours sitting around, idling away their time and often becoming extremely bored. It is no wonder that many of them, through luck or because of their own resourcefulness, were pioneers in finding gold.

Peters was resourceful. He was also cunning. One evening he decided to investigate a nearby outcrop of quartz rock. When he hacked pieces out of it, he was delighted to see in them the tell-tale glitter of gold. Unlike Jemmy Irving in New South Wales, Peters did not tell his employer, a Mr Barker, about his find. But he did share his secret with three of his workmates, and all four resigned promptly as employees of Barker.

They built themselves a crude hut in the forest and continued, successfully, their search for gold in rocky outcrops in the area. They found lots of gold, and would have found much more, if they had not been spotted by passers-by and reported to the authorities. Before long, the news of their find had been published in the *Argus*, the colony of Victoria's leading newspaper, and the area was ready for an invasion. Fortunately for Peters and his mates, the first report in the *Argus* gave the wrong location for the new goldfield, which gave them a little more time with the place to themselves. We do not know how much gold they found, but we do know that they sold a large quantity to a Melbourne business house. Eventually they were forced to pay a fine for mining without a licence, but with what they had earned, they had little trouble paying it.

A shepherd at work.

What miners found, when they arrived on the new goldfield, was abundant gold within a metre or two of the surface, spread over a large area. For many it was easy pickings. For others, inevitably, there

Early arrivals at Mount Alexander.

was nothing. But the new field was certainly the richest alluvial field yet found. Within a few weeks of the first arrivals, miners' tents and those of the butchers, blacksmiths and other merchants who eventually followed the diggers covered an area of about 40 square kilometres.

The new field became known as Forest Creek, but a couple of years later the government changed the name to Castlemaine. The nearby town of Castlemaine grew up as a supply centre for the diggings, which lasted about 10 years, until the abundant alluvial gold eventually ran out. There was little gold deeper down at the Castlemaine field. Unlike those at Ballarat and Bendigo, which were richer in reef gold—gold embedded in quartz rock deep down in the earth—the Castlemaine field had a relatively short life. It is estimated that by 1861, more than 14 million pounds worth of gold had been extracted from this particular Eldorado. At its height in the late 1850s, the population of the Castlemaine goldfield reached more than 30 000.

The naming of Castlemaine

During the 1850s, gold commissioners were among the most unpopular officials in the colony of Victoria. They were so unpopular because it was their job to supervise the collection of licence money, as well as to weigh and put a value on the gold that diggers found. In 1852 William Henry Wright was appointed chief gold commissioner for all the Victorian goldfields. Wright was the nephew of an Irish aristocrat, Viscount Castlemaine, and it is generally assumed that it was he who named the area Castlemaine, after his uncle.

BENDIGO

The town of Bendigo is in central northern Victoria, about 20 kilometres north of Castlemaine. Today it is a handsome town, surrounded by rich agricultural and pastoral country, with some fine stone buildings, reminders of the days when it was the thriving centre of the richest of all the Victorian goldfields. A feature of Bendigo is a 20-metre-high tower, known as a 'poppet head', which sits above a mine shaft that plunges more than 400 metres deep into the earth. This is the Central Deborah mine, which right up until the 1950s was excavating gold from deeply embedded quartz reefs. It is now a key tourist attraction, and visitors can descend into the shaft and see some of the workings of the former mine. Another Bendigo mine, the Victoria mine, was more than three times as deep. When it was sunk, it was the world's deepest gold mine.

Both these mines were sunk in the early twentieth century. By this time the gold rushes, which brought people flocking in their thousands to forage on and near the surface, had long since ended. In their place were companies that spent huge sums of money on machinery. The men who did the mining were employees of these companies. They laboured deep underground in damp, dirty, dangerous and unhealthy conditions and worked the rock face with mechanical drills.

Early finds

Bendigo's gold rush started in earnest just before Christmas in 1851. The area was part of the Ravenswood sheep station which had been established 11 years earlier. On part of the station was a series of waterholes known as Bendigo's Creek. It was named after a shepherd, whose hut was in the area, and who claimed to be a prize fighter. We can't be sure whether he really was, or whether he was just boasting. At the time there was a famous English boxer called William 'Abednego' Thompson, who was nicknamed Bendigo. The shepherd, too, became known locally by the same name, and so the creek, and later the town, got its name, although for a long time it was known officially as Sandhurst.

There are several accounts of the first finds in the Bendigo area. One story tells how Margaret Kennedy, the wife of the stationmaster at Ravenswood, found gold along the creek and then went out with a woman friend and some shepherds and

An early poppet head above a shaft at Bendigo.

Bendigo in the early 1850s.

started prospecting, the group spending their nights in crude tents made of bed sheets. Another account tells of a shepherd, Ben Hall, who was shown a piece of rock with a 'speck' of gold in it by a William Johnson, a Sydney man who was visiting the station. Hall pretended not to be interested, but when Johnson left, he promptly went back and started looking in earnest. It is clear that a number of people made finds because by early December there were more than 200 people prospecting in the region.

On 13 December 1851, the news became public. On that day Henry Frencham, who was both a journalist and gold-seeker, broke the story of Bendigo gold in Melbourne's *Argus* newspaper. The rush started immediately. People from nearby Mount Alexander who had not been lucky, or who had worked out their claims, packed their belongings and moved north, and more hopefuls from Melbourne left their jobs and added even more to the problems of the colony's capital.

Problems

We noted earlier how businessmen in Melbourne and farmers on the land watched helplessly as their employees left them and made for the goldfields. Perhaps even more worrying was the fact that the police and soldiers were also deserting in large numbers. Ships' captains, too, found their vessels stranded in Port Phillip Bay as their crews succumbed to gold fever, getting ashore as best they could, and never returning to duty.

Governor LaTrobe and other senior officials expressed deep concern about a breakdown of law and order and even the collapse of the whole colony. As we will see, there were serious troubles and violent incidents to come. But what almost no-one foresaw was that within a few months people from all over the world would pour into Victoria and that a few years later Melbourne, thanks to the gold rushes, would become the largest and most prosperous city in any of the Australian colonies.

FURTHER AFIELD

In spite of the initial rush, the Bendigo goldfield did not really come into its own until May 1852, when a rich find was made at a place called Eaglehawk Gully. By that time the colony had begun to be overrun by waves of gold-hungry immigrants.

The period following the initial find, late December 1851 and January 1852, was the height of summer, and the area was in the grip of a drought. Many of the creeks in the Bendigo region had dried up and prospectors found panning for gold difficult in the dry conditions.

Many of the people who were looking for alluvial gold had little or no experience and very basic equipment. Some merely foraged along the edges of creeks, scooping the topsoil and immediate subsoil with simple tin dishes or spades, and relying on the creek water to wash away the dirt and separate out any small pieces of gold. Others had more sophisticated wooden 'cradles'. You can see them in some of the pictures in this book. A cradle consisted of a large wooden box, with one side opening out onto a tray and with a rough screen, called a 'hopper', on top. At the front of the box, at the edge of the tray, was a small rise, or 'shelf'. Soil, sand and gravel gathered from the creek's edge were mixed with water and poured into the hopper, which filtered out the heavier stone and clods of earth. The finer material fell to the bottom, and the cradle was held tilted gently forwards and rocked carefully to allow the dirt to wash away. The remaining sand and, if the digger was lucky, some grains of gold, were then trapped behind the shelf. Men often worked in groups of three or four. One might dig the soil and gravel, another mix it with water and a third work the cradle. Only the most persistent of diggers, or those prepared to try their luck deeper down, could hope to succeed in drought conditions. Some gave up; others moved on to other fields where water was more abundant.

Antoine Fauchery's self-portrait.

Sometimes there were small gold rushes to areas where it was known that gold had been found. As you can see from the map on page 26, over the next 10 years, but especially during 1852 and 1853, the quest for gold spread widely throughout Victoria, reaching north almost to the New South Wales border.

Lucky and less lucky fields

During the 1850s, often called Victoria's 'Golden Decade', Ballarat and Bendigo yielded the greatest riches. Diggers who were prepared to put in the effort needed to dig deep in the earth streamed back to the almost abandoned Ballarat diggings during 1852, and the numbers swelled much further in 1853 when a number of large nuggets were unearthed almost 20 metres down.

Other, more remote, places had mixed fortunes. The Ovens field, in the far north of the colony, around which the town of Beechworth grew, provided rich pickings indeed for many. In early 1852, a former shepherd found gold at a place called Spring Creek. By the end of the year, about 8000 people had trekked through mountainous country to prospect there. Five years later,that population had almost trebled. The rushes continued, and by 1866 almost 130 tonnes of gold had been extracted from the area.

Another remote goldfield, at Omeo in the Gippsland mountains, had a different story.

Omeo was already an established settlement, a stopping place on an overland stock route, when gold was discovered in the area in 1852. A few hundred diggers, some of them from the Beechworth field, braved the cold to try their luck there, but few had any success. Some of them died trying to cross the mountains back to Beechworth.

The Mount Korong field near Wedderburn was another disappointment. A Scotsman found gold here in 1852, attracting several thousand hopeful diggers, many of them Californians and workers from nearby pastoral stations. Although some large nuggets, mined from shafts sunk into the ground, were discovered, this field could not compare with Bendigo and other rich fields to the south-east, and soon dwindled.

Striking it rich

While many people made great fortunes on the Victorian goldfields, no-one did better than George Lansell. Coming from humble beginnings, this adventurous businessman made and spent millions of pounds in developing mines in Bendigo during the 1870s. He constantly risked huge sums of money sinking deeper and deeper shafts in the Bendigo area, always confident that eventually he would strike gold. He almost always did.

Lansell was born and grew up in the town of Margate, Kent, in England. He worked as a candle-maker and grocer in Margate until he came to Victoria in 1853, at the age of thirty. He worked successfully for a time at his old trade and also bought and sold corn. When he first started investing in mining companies, he lost money, but he paid experts to advise him and his ventures began to prosper. Not content to retire comfortably on the modest fortune he had already made in the early 1870s, he kept investing in new and bigger mines. One mine in which he invested was called 'Lansell's 180'. Within weeks of buying it, he had made a profit of £180 000—considerably more than the mine had cost him to buy. Eventually, in 1890, the '180' mine would reach a depth of more than 800 metres. In one of Lansell's mines, gold was found at more than 1250 metres beneath the surface.

Lansell was a world away from the typical diggers who had trudged to the goldfields two decades earlier. He loved to visit his mines and

Unearthing the Welcome Stranger nugget.

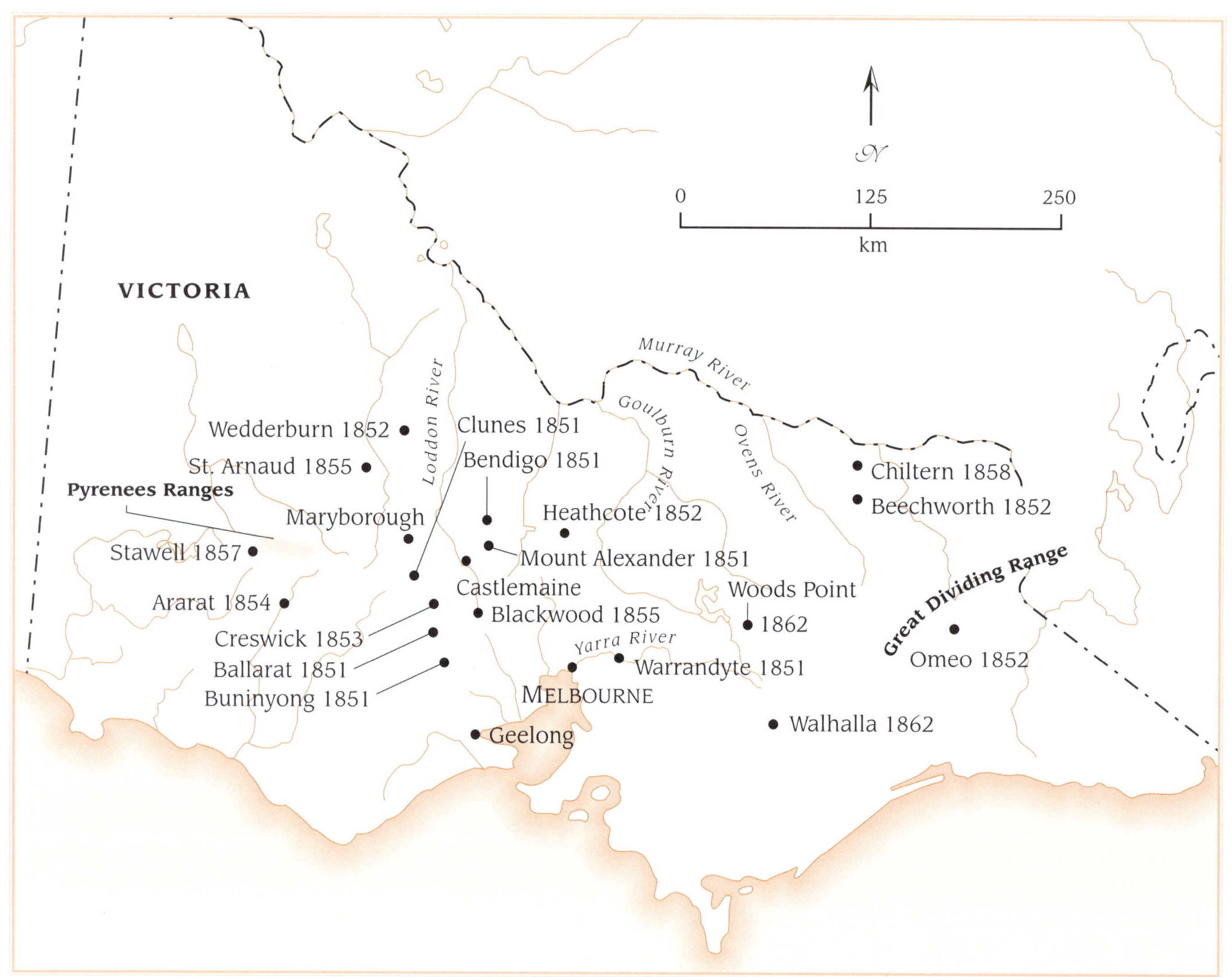

Victorian goldfields.

to survey the deep holes in the ground. But he rarely went down the shafts, preferring to issue his orders from the relative safety and comfort of the surface.

Many others profited from Lansell's success. Thousands bought and did well from shares in the mines he opened and thousands more found employment working in them. Except for a seven-year period, when he returned to England, Lansell spent the rest of his life in Bendigo, where he was a highly respected figure. He built himself a fabulous four-storey mansion in Bendigo and lived in it until he died in 1906.

A photographer on the fields

Many of our impressions of the Victorian goldfields come from Antoine Fauchery, who came to Melbourne from France in 1852. Fauchery was a pioneer photographer who was also a digger. He travelled around the Victorian diggings over a period of almost 20 years, taking photographs and writing copious letters, in which he gave detailed descriptions of goldfields life. In some of these letters he vividly describes the difficulties and dangers of working deep underground on the Ballarat goldfields. In 1854 he worked for seven months, with eight partners, on a shaft at Ballarat. In one terrible week during that time, Fauchery describes how seven miners in their area died in underground accidents. Fauchery and his partners found very little gold and eventually Fauchery sold his share in the venture for a mere £60. At this point, more than two years after he had started out,

Fauchery complained in a letter that he was 'hardly any further forward than on the first day'. Eventually, however, Fauchery did prosper in Victoria.

One of his most famous pictures is on page 25. It shows a group of miners from Cornwall in England supposedly unearthing the largest gold nugget ever found in Victoria, or in any of the Australian colonies. Known as the Welcome Stranger nugget, it was discovered at Moliagul, west of Bendigo, in 1869, long after most alluvial gold in the colony had been found. The nugget, which was luckily exposed to view by the wheel of a passing cart, weighed more than 78 000 grams, of which 71 000 grams was pure gold.

Prospecting for alluvial gold.

New South Wales Again

Following the initial rushes to Ophir and the Turon there were no further gold rushes in New South Wales until the early 1860s, when there was again a brief period of feverish activity. Further important finds at the beginning of the 1870s kept gold fever alive in New South Wales for an additional brief period.

Victoria was undoubtedly the golden colony during the 1850s and 1860s, when its diggings attracted worldwide attention. In the 1850s, its goldfields produced more than 14 times as much gold as those in New South Wales. For some time, most of this gold was found on or near the surface, but by the middle of the decade, most of the alluvial gold in the main goldfields had been found and miners had to resort to more sophisticated methods to extract reef gold from deep below the surface. This made prospecting not only more difficult, but also much more dangerous. As Antoine Fauchery's experience shows (see page 26), lives were lost and injuries sustained as men laboured deep in the earth, often chipping away at rock faces in poorly constructed galleries, supported by flimsy planks of wood or pieces of tree trunks.

Even for miners whose shafts did not go deep, windlasses, which lowered buckets into shafts in the ground, replaced the simpler dish, shovel and cradle as the miner's principal tools. Often mechanical pumps were needed to pump excess water out of the shafts. Under these circumstances, miners generally had to work in teams of three or four, rather than alone.

Reef mining at Ballarat, 1855.

These, too, were the conditions under which most people worked on the New South Wales goldfields of the 1870s. Some found payable alluvial gold, but most of the riches lay hidden away below. By this time, too, mining companies, or syndicates of miners, which could afford to invest in expensive mining equipment, were major players in the search for golden riches.

Further New South Wales finds

The first new New South Wales field did not amount to much. Late in 1859 people on a sheep station found gold high in the Snowy Mountains at a place called Gibson's Plains. By April 1860 there were 10 000 people there prospecting for gold and the town of Kiandra was established. A year later only about 200 hardy individuals remained. The rest had been driven away by the cold winds and snow of the alpine winter, which had arrived hard on the heels of the rush to the area.

In the 1830s, Lambing Flat was established as a sheep run on the plains of southern New South Wales. In March 1860 there was an American cook at Lambing Flat. He was known as 'Alexander the Yank' and he had been on the Californian goldfields more than 10 years earlier. It occurred to him that the quartz rock at Lambing Flat was very similar to the rocks he had seen in California. He went prospecting in a creek with a billycan lid and found gold. As usual the word got out, and in little less than a month there were 400 miners working the area. By the summer there were ten times that many, including a large number of Chinese gold-seekers. Many of these, like others who came to the new fields, had come north from Victoria. Also among those at Lambing Flat in 1861 were Ben Hall and Frank Gardiner, who were to become two of the most notorious of Australian bushrangers. Gardiner, who already had a long criminal record, sold stolen meat to the diggers. Hall, too, was a butcher at the diggings. In 1861 the town that grew up around the diggings was named Young, after Sir John Young, the new governor of New South Wales.

Frank Gardiner.

Further north, the town of Forbes was established after gold was discovered on a sheep run there in the middle of 1861. There were 120 people in the area when gold was discovered. A year later there were about 30 000. But the alluvial gold soon ran out, and only the more energetic, and better equipped, miners, who were prepared to dig deeply buried quartz reefs, stayed on. In 1863 a mere 3500 miners remained. But mining companies, which had the machines and employed the manpower to exploit the deep deposits, moved into the area and continued operations for another 50 years.

Rushes of the 1870s

In 1871 the famous English novelist Anthony Trollope travelled widely around New South Wales. When he arrived at Gulgong, he found a town of about 12 000 people, the centre of a thriving goldfield. Along the way he saw groups of men trudging towards the town.

A gold rush was on, but, according to Trollope, the men 'seemed to me to rush very leisurely'. Although Gulgong's buildings were built crudely of slabs of wood—Trollope wrote that each one 'had probably required but a few days for its erection'—there were all kinds of merchants plying their trade. He noted 'bakers, butchers, grocers ... dealers in soft goods ... an auctioneer's establishment ... a photographer and ... a theatre, at which I saw [a play] acted with a great deal of spirit'. Yet, barely a year before, Gulgong had not existed. Although gold had been found nearby 20 years earlier, there had been no rush to the area, as there had to the Turon fields about 80 kilometres to the south. It was not until April 1870, when a man called Tom Saunders found gold valued at £120 at a place called Red Hill, that a gold rush began. Two other smaller towns, Home Rule and Pipeclay, also grew up around these new goldfields, which had, however, been mined out by the mid-1870s.

Gulgong today is a sleepy little town of almost 2000 people. Many of the buildings which Trollope saw and described, including the theatre, are still standing and are beautifully preserved. In an old bakery is a museum, the Pioneers' Museum, where you can see photographs and other relics of Gulgong's gold rush days. There is even a miner's cottage,

Miners in Gulgong, 1870s.

complete with original furniture, and an old schoolroom containing many photographs of school life on the goldfields.

Hill End on the Turon is another story. It is now almost a ghost town, with one street, and a few straggly buildings. It is a far cry from the booming town of the early 1870s, which boasted four churches, three banks, a school, two newspapers and no less than 50 hotels. One of the hotels, the Royal Hotel, still stands, and still takes in visitors. The old hospital, too, has survived and is now a museum. In the area around the town you can see the deep shafts which individuals and about 200 syndicates and mining companies sank in order to reach the gold-rich reefs. Hill End had been part of the first Turon rush of 1851, but had not really thrived. Alluvial gold could be found much more easily at nearby Tambaroora. But by 1872 Hill End had very much come into its own.

Holtermann with his 'nugget'.

The town attracted both newspaper headlines and thousands of curious visitors in late 1872. On 19 October, deep in a shaft on their claim at Hill End, German immigrant Bernard Otto Holtermann and his business partner, Louis Beyers, found a huge piece of quartz rock that weighed almost 300 kilograms and contained gold worth £12 000. It was the largest piece of reef gold ever discovered and although it was not, strictly speaking, a nugget, it became known everywhere as 'Holtermann's nugget'.

Holtermann and Beyers had been prospecting in the area, without much luck, for several years when, some time in 1871, they struck payable gold at Hill End. They formed a company, the Star of Hope Gold Mining Co., and were already rich men when they made their most spectacular find. Holtermann, who was a pioneer in the still relatively new craft of photography, had himself photographed standing beside his nugget. He also commissioned two other photographers to travel around the Victorian and New South Wales goldfields and record scenes there. This great collection, known as the Holtermann Collection, is an important source of information about the goldfields and gold towns of the 1870s.

NORTHERN GOLDFIELDS

We have already read on page 29 of the bushranger, Frank Gardiner. He was at Lambing Flat, now Young, in 1861, working as a cattle thief and supplier of stolen meat to the diggings. In 1862, he masterminded a spectacular gold robbery near Forbes, further to the north. Soon after, as the law closed in on him, Gardiner fled to Queensland, where he ended up at gold diggings at Appis Creek, about 100 kilometres north-west of the town of Rockhampton. After living there for 18 months, Gardiner was captured in 1864 by three policemen who were disguised as gold miners.

Although gold was being mined in Queensland in the early 1860s there were no major goldfields there until 1867. At that time Queensland had been a separate colony for only eight years. The young colony depended heavily on sheep and cattle-raising and agriculture. But by 1867 things had turned bad for Queensland's farmers and pastoralists as the country was in the grip of a vicious drought. So when, in October 1867, James Nash walked into the police station at Maryborough and reported that he had struck payable gold in the hills about 80 kilometres to the south-west, he sparked an immediate rush. Gold would be the saviour of Queensland's economy and would, during the next two decades, stimulate the northward expansion of

The Day Dawn mine at Charters Towers in the 1880s.

Early days in Gympie, Queensland, centre of the Nashville Diggings.

Hopeful diggers arriving in Queensland.

the colony's population and the establishment of several major towns.

Work was hard to come by in drought-stricken Queensland, and many men, like Nash, had taken to prospecting for gold. One night in September 1867, Nash camped with some timber-getters. One of them, who had been on the Victorian goldfields, told him of an area in the vicinity that looked to him like gold country. Nash took the man's advice and a few days later he found gold worth more than £200 in a creek in an isolated gully.

The gully did not remain isolated for long. Within weeks, hopeful miners on foot or in bullock-drawn carts had carved out rough roads over the hills, and a town of canvas tents and bark and slab huts grew up on the banks of the Mary River. At first it was known as Nashville, but in 1868 the name was changed to Gympie, after the Aboriginal name for a kind of stinging bush that grew abundantly in the surrounding region.

Within a year almost all the alluvial gold at Gympie, worth more than half a million pounds, had been removed. Then, as at other goldfields, groups of miners combined to dig and mine shallow shafts. Within 10 years, however, all the gold close to the surface had been mined and attempts to dig deeper were foiled when miners hit a mass of hard greenstone which they could not penetrate. That could have spelt the end of Gympie's golden days, but in the 1880s mining companies with sophisticated equipment drilled through this barrier to the gold-bearing quartz below. Gympie remained a gold town for another 40 years. By the time the gold was exhausted in the mid-1920s, this Eldorado had given up almost 120 tonnes of gold.

Further north

Townsville is an important port city in north Queensland and is Australia's largest tropical city. In 1868, when gold was first discovered at Ravenswood, about a hundred kilometres to the south-west, it was a tiny, struggling settlement. But the next year it received a boost when it became the gateway to the new diggings. For a while Ravenswood thrived on the alluvial gold that was found there. In 1871 there were about 3000 miners in the area, and 42 hotels, some of them no more than tents or slab huts, helped quench their thirst. Once the alluvial gold ran out and shafts were dug, Ravenswood's troubles began. When they reached the water table, miners found the gold mixed with other metals, such as lead and zinc, and they had no way of separating it out.

At the end of 1871, a summer storm and a team of frightened horses led to the discovery of what was destined to become Queensland's greatest goldfield. Just before Christmas that year, Hugh Mosman and two other gold diggers from Ravenswood were prospecting on pastoral land about 65 kilometres to the north-west. A storm blew up and their packhorses panicked and ran away. A 10-year-old Aboriginal boy called Jupiter, who was with Mosman and his companions, went to round them up and found gold in a nearby creek. To the struggling miners of Ravenswood, Mosman's news came as a godsend. Most of them immediately left for the new diggings, which became known first as Charters Tors, and later as Charters Towers. Many more soon joined them and the new field went from strength to strength. In 1873 there were 4300 prospectors there. Thirteen years later, there were more than 11 000 and at the turn of the century there were almost 30 000. Commercial mining companies were quick to set up in Charters Towers and in the mid-1880s there were more than 150 mines in the area and the landscape was littered with their poppet heads, standing over shafts that descended almost a kilometre into the earth. One of these mines, the Day Dawn, was sold to a London company in 1887 for £640 000.

In spite of Jupiter's lucky find, there was very little alluvial gold at the new field. The early prospectors had to work hard for their gold, which is one reason why the Queensland fields never attracted the numbers that those in Victoria did. Another, and more important, one was their remoteness. A third, and perhaps most important of all, was the climate. If miners froze in the winter at Kiandra and at some Victorian fields, they cooked year-round in the unrelenting tropical heat of Charters Towers and even more so in the fields that were opened up still further north.

Queensland goldfields.

Near the tip of Cape York

Queensland's Palmer River, which rises in the Great Dividing Range and flows westwards across the north of Cape York, was unknown to white settlers until an expedition led by William Hann reached and named it in 1872. The following year the explorer and gold prospector James Venture Mulligan found payable gold along its banks and a rush followed. The nearby town of Cooktown was established as a port to serve the new diggings, which grew up around a settlement called Maytown. At its height in 1877 there were almost 20 000 miners in the area. The great majority of them were Chinese.

In 1874 Mulligan discovered and named the Hodgkinson River, which rises in the mountains just inland from Cairns. The following year, he was probably the first to find payable gold here too. A rush followed, but there was little alluvial gold, and reef mining, given the difficult conditions, proved very hard work and was often unprofitable. In 1878, many of the miners left to join a small, but not very successful, rush to a field at Coen, even further north.

These north Queensland fields saw some of the most violent scenes in Australia's gold rush history. In these places, Chinese diggers greatly outnumbered Europeans. Some estimates put the numbers of Chinese on these far northern fields at about 20 000. This led to racial tensions which, in the torrid conditions, often erupted into clashes. There were also clashes between different groups of Chinese. In 1878 at a place called Lukinville on the Palmer River, warfare broke out between rival groups of Chinese from Macao and Canton.

The worst violence, however, involved the local Aborigines. Displaced from their hunting grounds by the new arrivals, they took their revenge by spearing both European and Chinese invaders. The miners responded in kind, except that their guns proved more effective

Cooktown in the late 1880s.

weapons, and many of the local place names, such as 'Hell's Gate' on the Palmer, reflected the bloody encounters that occurred there.

The experiences of Thorvald Peter Ludwig Weitemeyer, a Danish visitor to the Palmer, give some idea of the difficulties and dangers of prospecting in this area. He described how visitors to the Palmer needed to carry months of provisions with them or risk starving to death. Without the hunting skills and local knowledge of the Aborigines, these overland trekkers found the country 'very poorly off for game'. Many had to resort to killing and eating their horses and dogs. He tells the sad story of one man who was isolated by a flooded river and died from hunger in his tent. He was found later after the rains had subsided. His head was lying on a bag. This bag contained a fortune in gold.

Diseases, too, were rampant in the hot and humid climate. Dysentery was very common and caused many deaths. Those who suffered from it could expect little sympathy from other miners. Even in Cooktown, where he found thousands of people 'camped out in tents', Weitemeyer observed men 'lying helpless, writhing with pain on the ground, some of them bellowing out', but receiving no 'pity or mercy' from passers-by.

South Australia

There were no major gold rushes in what we now know as South Australia, although there was some excitement in 1852, when William Chapman discovered payable gold at Echunga, about 30 kilometres south-east of Adelaide. About 4000 hopefuls rushed to the area, but the pickings were mostly meagre and the gold soon ran out. At the time South Australia was enjoying a mining boom of another kind. Copper had been discovered in the south- east of the colony in the 1840s and the mines that had been established certainly helped stimulate an interest in finding minerals, especially gold, in New South Wales and Victoria. Further finds in the desert areas, around Tarcoola, further to the north-west, brought more miners to South Australia during the 1890s.

In 1863 what is now the Northern Territory was transferred from New South Wales' control and became part of the colony of South Australia. Nine years later, a worker, D'Arcy Wentworth Ur, who was part of a team digging holes for the telegraph poles that would carry the Overland Telegraph between Darwin and Adelaide, and link the Australian colonies by cable with the rest of the world, discovered gold on the banks of Pine Creek. There was an immediate rush to the area, but the extreme remoteness of the region and the difficulty of getting there ensured that this rush would not be a small one. The main players at Pine Creek were destined to be,not small prospectors, but mining companies. They moved equipment overland from the port of Darwin. But when they found it difficult to entice Europeans to come and work in this remote region—the continent's most northerly goldfield—the management of the 15 or so mines that set up in the area actively encouraged Chinese migrants to come and work there. There was an advantage in this, because a Chinese worker would work for about half as much money as a European.

In 1889 a railway, also built mainly by Chinese labour, linked the mines at Pine Creek to Darwin. In 1894, when the mines at Pine Creek produced almost one and a quarter million grams of gold, 97 per cent of the 2100 or so workers there were Chinese. As well, Chinese people also played a leading part in the management of about half the mines, and part-owned some.

Several of the mining companies that came to Pine Creek did not prosper because it proved difficult and expensive to extract the gold from the quartz. Most mining there closed down at the end of the nineteenth century. However in the mid-1980s gold mining returned again to Pine Creek in the form of a large open-cut mine. Once again the landscape of this tiny settlement was transformed, but instead of the poppet heads and shafts of a century earlier, there would be one deep gash in the earth and towering piles of waste materials.

NEW ARRIVALS

Australia is often called a land of immigrants. In modern times the country's population has been dramatically increased several times, and substantially changed, by large 'waves' of new migrants. We now know that Aboriginal people had lived in all parts of the continent and Tasmania for probably more than 40 000 years before 1788, when Britain sent the 11 ships of the First Fleet, carrying almost 1500 convicts, soldiers and officials to establish a penal colony in New South Wales. By 1850, just before gold was discovered, more than 330 000 immigrants had arrived in the colonies from Britain and Ireland. A little over half of these were free immigrants; the rest were convicts. The 1830s had seen a large increase in free immigrant numbers as colonial governments offered assisted passages to encourage new settlers, especially young women, who were in short supply, to come to live in the colonies.

The number of migrants dwindled in the 1840s, as the colonies suffered from a severe economic downturn, but with the discovery of gold, there was a huge surge. Between 1851 and 1861, the total population of the colonies more than doubled, from 405 000 in 1851 to 1 145 580 in 1861. The population of New South Wales doubled; Victoria's population increased seven times over.

More than 600 000 people sailed to Australian shores from overseas countries in this time. More than four out of every five of these came from Britain and Ireland, and a little over one-third of these came on assisted passages. It has been estimated that one out of every 50 persons in Britain and Ireland sailed to the colonies over the decade. About 8000 crossed the Tasman Sea from New Zealand, and more than 10 000 made the longer voyage across the Pacific from the now-exhausted Californian goldfields. Some of these were Americans, but the majority were miners from Europe and the Australian colonies who had failed to find their fortune in California. Of the non-British people who came to the colonies in that time, the greatest number were from China.

The number of Chinese people in the colonies varied, as new migrants arrived and others sailed for home. By 1854, there were only 2000, but by the following year there were five times that many. In 1861, there were just under 40 000 Chinese in the colonies, the great majority of them on the Victorian goldfields. Of these only 11 were women. This contrasted sharply with the English and European migrants. About one in every three of these was a woman, and many of these women worked, alone or with their men, on the goldfields.

A slow start

Most people in England in 1851 knew almost nothing about the Australian colonies. If they ever thought about them at all, it was probably as places inhabited mainly by convicts and strange animals, where the climate was harsh and the landscape scrubby and uninviting. When the first reports of gold in the colonies were published in English newspapers late in 1851, few took much notice. Many dismissed them as either false or greatly exaggerated. But as the reports persisted, and especially those of fabulous finds at Mount Alexander, interest increased. It reached fever pitch when, in the middle of 1852, six ships from Victoria arrived, bringing eight tonnes of gold. The Australian colonies were the talk of London and of many other towns, as thousands hurried to get passages on southward-bound ships.

All sorts and classes of people came, though the majority were young men. Many were educated people who forsook promising careers in Britain for the uncertainties of goldfield life. The famous writer Charles Dickens caught something of the spirit of the times when he wrote that 'legions of bankers' clerks, merchants' lads ... secretaries, and ... cashiers,

The Marco Polo.

all going with a rush, and all possessing [only] confused ideas of where they are going and what they are going to do'.

But Dickens was only partly right. Many of those who sailed were experienced coal and lead miners, who would have had few illusions about the difficulties and dangers of mining. Certainly what attracted many were figures published in various guides, giving more or less accurate information about prices and wages for different jobs in the colonies. Even if they did not strike it rich on the goldfields, many people reasoned, they could probably earn more and live better than in Britain.

As it turned out, a lot of gold-rush migrants never went to the goldfields, or went there only briefly. Many of them found jobs in Melbourne and so contributed to the prosperity and growth of that city, which at first had been threatened by the gold rushes. Others went to the diggings and provided in various ways, as merchants or as tradespeople, for the needs of the diggers. Many helped to establish the towns that grew up around the goldfields.

The voyage

The goldfields were often rough and dangerous places, but the voyage from England was usually rougher and more dangerous. Many a would-be digger never made it to the diggings, but ended up buried at sea or wrecked on the Australian or some other foreign coast. The great numbers wanting to emigrate meant that some old, unseaworthy ships made the trip, and conditions on board were often hopelessly overcrowded.

Even worse than the threat of shipwreck was that of disease. During 1852, about one in 20 people making the long voyage died on the way. The journey could take up to five months, the food was often poor and the weather, both above and below decks, was sometimes freezing. A new clipper, the *Marco Polo*, made the journey in the fast time of 68 days in 1852. When it arrived in Melbourne, 53 out of its 930 passengers had died. Also in 1852, typhoid fever broke out on the *Ticonderoga*, killing 165 of its passengers. The following year, one ship from Liverpool lost 101 passengers from fever and was quarantined before anyone was allowed to land.

English migrants arriving in Melbourne.

The ships that set out from European ports and from California were, by and large, worse. On one ship that arrived from California in May 1853, the passengers spent most of the time pumping water out of the leaking hold. When they complained, the captain informed them that they had two choices: they could pump, or they could swim the rest of the journey.

But the worst conditions were suffered by the Chinese emigrants, many of whom made the journey below decks in unspeakable filth. The great majority of these Chinese diggers were poor peasants who could not afford to pay for a passage. Their fare was paid by Chinese or other foreign businessmen who made money out of the Australian gold rushes without ever going near the diggings. In return for their passage, the Chinese diggers were forced to pay a part of the gold they found. The rest, if any, was to help them and their families escape from the poverty in which they lived.

Arriving and disembarking

People who survived the voyage were usually greeted in Melbourne by scenes of chaos and confusion. Some were reminded of similar scenes when they had arrived in San Francisco some years earlier. To begin with there was more often than not a queue of vessels outside Port Phillip Bay, waiting to be guided in. In May 1853, 69 ships from overseas arrived in Melbourne to join the hundreds of others that littered the bay, many of them unable to leave because their crews had abandoned them for the diggings.

It could be another two days before they could disembark, and passengers were often so disconcerted by how much it would cost to have themselves and their belongings taken ashore, that they simply heaved their luggage overboard. Then there was the problem of finding accommodation. If they knew no-one in the town, they often took up residence at first in the spreading mass of tents known as

Canvas Town that grew up in 1852 on the southern side of the river. Here, for five shillings a week, a new arrival could sleep under cover, even if they had to lie on the bare, damp ground. Others who could not, or chose not to, pay simply slept in the open on the wharves. Those who slept outside often made the wiser choice, for Canvas Town was a grim place. By the end of 1852 about 7000 people were living there, many of them sick or dying.

Ellen Clacy, who arrived in 1852, and who described her experiences in a book, told the following 'sad but common tale'. One day she found a small child carrying water to a tent. She followed the child and found inside a sick woman and another child. Mrs Clacy came back next day with food, but the woman had died overnight. A day later, the woman's husband, who had gone on alone to the diggings, returned with payable gold.

The following year the government closed down Canvas Town and moved it to another location—not because it was unhealthy, but because of the criminal types who were congregating in the area.

A frontier town

A Melbourne shopkeeper who opened a general store in 1851 described his fellow townsfolk as 'the sober, plodding and industrious people of Melbourne'. That was before the gold rushes. The people who arrived from 1852 onwards found a very different sort of place inhabited by a very different kind of people. Many must have wondered and worried at what they saw, because the place now had more the air of a frontier town than of a capital city. For many, it was a reminder of all the worst things they had heard or read of the wild and lawless San Francisco of the late 1840s.

Many locals who had not left for the diggings decided that there was easy money to be made from the new arrivals. Not only did the men who rowed out to meet the incoming ships charge exorbitant amounts of money to bring people and their belongings ashore, they also refused to give the newcomers full value for the English bank notes which was the only money the new arrivals had to pay them with. Those who stayed for any time in Melbourne saw swaggering diggers who had struck it rich either lounging about the street or racing their horses up and down, churning up the dust. Others, keen to show off their new wealth, paraded or raced through the town in elaborate and expensive carriages. Pubs, where the newly rich could easily and quickly unload their earnings, abounded in the town.

Riotous behaviour was the rule rather than the exception as fortunes that were hard won at the diggings were rapidly squandered. One observer called Melbourne a 'most frightful place for drinking'. Another summed the place up neatly by saying that 'diggerdom' was 'gloriously in the ascendant' there.

Digger's wedding in Melbourne.

TROUBLES ON THE GOLDFIELDS

There is little doubt that at least some of the men who sported their new-found wealth so publicly in Melbourne had won their money dishonestly. While it seems that by and large diggers were honest and industrious, wanting only to improve their lot inlife, the diggings inevitably attracted a criminal element. In the early days of the gold rushes especially, the roads and tracks to and from the diggings were dangerous places as many men took to bushranging as the easiest way to make a living. Even close to Melbourne the roads were not safe. An English writer, William Howitt, who went to the Victorian diggings in 1852, described an incident in October of that year, when four armed men robbed 20 separate travellers on the St Kilda Road in broad daylight and then made off into the country, 'bailing up ... everyone they met'.

Ben Hall.

Most people almost automatically blamed ex-convicts from Van Diemen's land for the troubles in Melbourne and at the diggings. Ellen Clacy reflected the general mood when she wrote that about one-tenth of diggers were 'composed of outcasts and transports—the refuse of Van Diemen's Land—men of the most depraved and abandoned characters'. As we noted earlier, attempts to keep these people out of Victoria had failed miserably.

Another group that was widely suspected of stirring up trouble were Californians. Reports, largely accurate, about lawlessness at the Californian diggings were widely circulated. A young English visitor to the Mount Alexander diggings who would later be four times prime minister of England was Lord Robert Cecil. With perhaps a hint of snobbery, he described a Californian fellow passenger on the coach to the diggings as a 'coarse, hideous, dirty-looking man ... [who] wore ... a pair of earrings about the size and shape of a wedding ring ... [and] a pair of pistols in his belt'. Even worse, 'the words "put a bullet through his brain" were constantly in his mouth'. In spite of this unpromising introduction to the Victorian goldfields, Cecil concluded that the diggings were relatively peaceful places where there was 'less crime than in a large English town'.

A government official at the Bendigo diggings was in no doubt about the bad influence of Californians. He wrote of 'adventurers ... from [the] California diggings, who were inciting people to set up Judge Lynch'.

Rough justice

Despite Robert Cecil's judgement, the diggings were often wild and violent places and lynchings, though not common, did occur. A crowd of angry miners at Mount Alexander in 1852 promptly hanged a miner who had killed another miner after an argument. Another, who had been caught trying to rob another miner's tent, was lucky to escape with a severe flogging after an angry mob were threatening

Miners come to blows in a dispute about a claim.

to hang, shoot or drown him. As the crowd argued about the appropriate punishment, a commanding giant of a man, Bob Raikes, came upon the scene, calmed the crowd and managed, at least, to save the culprit's life. In another incident at Ballarat an incensed mob held a would-be thief over afire, almost cooking him alive.

Guns, arguments and grog

The Californian described by Robert Cecil was not alone in carrying firearms. In fact the miner who did not have a gun was more likely to be the exception. Antoine Fauchery wrote that 'everyone [on the goldfields] is armed to the teeth, and if one really does run any risk, it is not from bushrangers ... There are enormous chances of being killed by one's neighbour'. Ellen Clacy painted a vivid word picture of a night at the diggings: 'Imagine hundreds of revolvers almost simultaneously fired ... murder here—murder there—revolvers cracking—blunderbusses bombing—rifles going off—balls whistling ...' Even allowing for some exaggeration for dramatic effect, it is clear that the diggings, atleast before an effective police force had been formed, were dangerous places. Many a dispute that began with shouting and then progressed to a fist fight ended with a gunshot.

There were many things to fight about on the goldfields. There were disputes about claims, jealousies about neighbours' lucky finds, allegations of thefts and fights over women. There was also the sheer frustration of working hard in uncomfortable conditions, often for little or no reward. Men, long separated from women and children, were very

Bushrangers of the goldfields.

likely to crack under the strain. They were all the more likely to resort to violence when they had had too much to drink, and some of them regularly got drunk, usually on whisky, rum and other spirits.

Alcohol was freely available on the New South Wales fields and makeshift pubs were abundant and constantly patronised. In Victoria, until 1853, the government banned all alcohol on the goldfields. As a result 'sly grog' tents or rough slab huts, often pretending to be coffee shops and often run by women, were a feature of every goldfield and there were many scenes of drunken revelry and fighting. Even when police were present, they turned a blind eye to these goings-on. They knew that they had no chance of stopping them. Besides, many, perhaps most, of them happily took bribes to 'look the other way'. Grog was 'smuggled' to the goldfields, hidden on large carts beneath other provisions. If a cart was stopped and searched, the driver had usually taken the precaution of obtaining documents to 'prove' that the liquor was bound for non-goldmining towns.

If Saturday nights were often riotous on Victorian goldfields, Sundays were usually very quiet, a fact that was remarked upon approvingly by many visitors. Often the loudest sounds to be heard were those made by the many preachers who delivered sermons to the surprisingly large crowds that gathered. Mining on Sundays was banned by law and miners almost invariably observed the ban.

Bushranging on a grand scale

Many of the bushrangers who haunted the goldfields were small-time criminals who robbed travellers of their possessions, money and small quantities of gold. Others, though, were thoroughly professional criminals, often at the head of organised gangs, who set their sights on the banks that grew up in the gold-rush towns, and on the gold escorts—heavily guarded coaches that transported gold from the diggings to regional centres and the capital cities.

One of the most daring robberies was the hold-up of the escort bringing gold from the McIvor goldfield, south-east of Bendigo, to the town of Kyneton. In July 1853 a gang of five armed and masked men ambushed the coach near a bend in the road. They shot the driver of the coach dead and injured

A gold escort near Bendigo.

the three troopers who were guarding it. They escaped with two boxes, containing gold worth £5000. Four of them ended up paying a terrible price for their crime. They were arrested some months later and three of them were hanged. One committed suicide. The last of the five, Jeremiah Murphy, was allowed to go free in return for informing on his mates.

New South Wales goldfields were free of bushranging on this scale in the 1850s, but in June 1862 a gang of eight, led by Frank Gardiner and including Ben Hall, ambushed the gold escort from Forbes at a place called Eugowra Rocks. The bushrangers peppered the coach with rifle fire, wounding two of the guards, but almost miraculously not killing anyone. The driver of the coach narrowly escaped injury or death as a bullet dislodged his hat. In the biggest gold robbery ever carried out in Australia, the gang escaped with gold to the value of £14 000.

Almost three years later, in March 1865, a gang led by Ben Hall attempted to rob a gold escort near Araluen, in south-eastern New South Wales. Although they shot and wounded one of the four guards, the other guards opened fire on the bushrangers, who were forced to flee the scene empty-handed. By this time it was near the end of the line for Hall. Less than two months later he was hunted down and shot by police at his camp site near Forbes.

The robbery of the McIvor gold escort.

ANTI-CHINESE FEELINGS

There were many nationalities on the goldfields of the Australian colonies. The vast majority were either Australian-born or from parts of Britain, which then included Ireland. There were also, as we have seen, a number of Californians as well as groups from different parts of Europe. Often these groups formed separate communities on the goldfields. In 1855, for example, Antoine Fauchery wrote that there were reasonably small numbers of Russians, Spaniards, Dutchmen, Belgians and his fellow Frenchmen at the diggings, but that there was a community of about 3000, 'almost the total population of [the] Swiss canton [region] ... of Tessin' at the Jim Crow field at Daylesford in Victoria. They included women and children and had been 'settled' there for almost two years.

But there was one group of foreigners who attracted particular attention. These were the Chinese. As their numbers grew in the mid-1850s, the European miners became more suspicious of and hostile towards them. There were a number of reasons for this. To begin with, the Chinese were conspicuously different from other miners. Apart from their 'Asiatic' appearance, they wore what looked to Europeans to be very strange clothes. They generally walked to the diggings in large groups dressed in long, loose, knee-length shirts, floppy trousers, turned-up shoes and wide, pointed, straw hats. Many of them, too, had their hair plaited into long pigtails. They carried their belongings hanging from long bamboo poles, which they rested on their shoulders.

When they arrived and settled at the goldfields, they had little or nothing to do with other groups and they built strange-looking temples, or joss houses, where they burnt incense. A joss house, built during the 1850s, still stands in Bendigo. They spoke a strange and mysterious language. They hardly ever drank alcohol, but they smoked opium, which was not illegal in the colonies at the time, and they ate very different food from the white miners. They also gambled heavily among themselves. Perhaps worst of all in the eyes of white miners was the fact that there were no women with them. This led to the suspicion that, given half a chance, they would attack or steal white women.

Another source of annoyance was that they were industrious and hard-working and had little time for recreation. While very few Chinese dug trenches to mine underground reefs, they often descended into trenches that had been dug and abandoned by Europeans to find gold that the less patient Europeans had left behind. This really annoyed other miners. And when they had finished at the diggings, they took their gold and money back to China. This should not have angered the Europeans, as many of them, too, sailed home with their winnings. But it was one more grievance to add to the list of complaints.

A petition, signed by more than 400 New South Wales gold miners in 1858, and sent to the colony's government, complained of the Chinese, describing them as 'a very ineligible class of immigrants', accusing them of having 'general filthy habits that are repulsive to the feeling of the Christian population', of being 'heathens' and of gambling and 'purchasing their supplies' on Sundays, which was a Christian day of rest.

Clashes between Chinese and white gold diggers became more frequent as the numbers of Chinese increased. In 1855 the Victorian government passed laws to make things more difficult for them. But they were no more successful than the laws aimed at keeping out ex-convicts from Van Diemen's Land had been. The law imposed a tax of £10 on every Chinese person who entered the colony. As a result, shiploads of Chinese landed in South Australia and walked overland to the diggings. Many died during the long and arduous trek. Two years later a new tax of £1 per month was imposed on every Chinese in the colony.

At this stage only a few hundred Chinese had come to the New South Wales goldfields, but in 1858, as a result of the Victorian laws, more than 12 000 landed in that colony. Some of them went south to Victoria, but many made their way westwards to the New South Wales diggings. Within three years, there were three Chinese to every two European diggers on the New South Wales fields.

Riots

The worst acts of violence against Chinese diggers occurred in Victoria in 1857 and New South Wales in 1861. In May 1857, a group of Chinese travelling overland from South Australia to Bendigo were attacked by a mob at Ararat, where the Chinese had unexpectedly struck gold. The Chinese were beaten up, driven away and many of their possessions destroyed. At that time there were about 3000 Chinese on the Buckland River field near the Ovens River in north-eastern Victoria. Many of the 1000 or so white diggers on the field were Californians, and on 4 July, American Independence Day, an angry mob of 100 or so whites attacked the Chinese, driving them out of their camp, burning and looting their tents.

A Chinese joss-house.

At least three Chinese died in this attack. Three whites were injured when the Chinese later opened fire on them in retaliation. A troop of police, led by Robert O'Hara Burke, who was later to lead the ill-fated Burke and Wills expedition to the north of the continent, arrived on the scene two days later. They arrested a number of the ringleaders. All but four of them were acquitted by sympathetic juries when they came to trial. Those four received short prison sentences.

Lambing Flat

The most violent anti-Chinese riot of all occurred on the Lambing Flat (now Young) goldfield in southern New South Wales in June 1861. Trouble had been brewing there for some time. In November 1860 about 500 Chinese had been driven out of the area and their tents had been destroyed, and early in 1861 two Chinese had been killed and about a dozen others wounded in a series of armed attacks on the Chinese encampment by small groups of whites.

But the crowd that gathered on the morning of 30 June 1860 was no small group. It numbered well over 2000, many of whom were drunk. The ringleaders had, however, carefully planned the attack. They had made anti-Chinese banners, and even assembled a small band of musicians. They stirred up the miners in a place called Tipperary Gully, and led them, to the accompaniment of marching music, towards the Chinese camp at Back Creek. There was an ugly festivity in the air as the mob, singing, chanting and yelling, took to the unprepared Chinese with whatever weapons were to hand—pickaxes, spades and whips. Many had guns. More than a thousand Chinese fled the scene and took refuge at a nearby property, whose owner, James Robert, fed and sheltered them for several weeks. Others who were unable to get away, or who stayed behind to hide their gold and other possessions, were set upon brutally. The assailants battered them with their weapons and seized their pigtails and cut them off or in many cases ripped them out of their scalps.

Chinese miners on their way to the diggings.

Police confront rioting miners at Lambing Flat.

According to one report in a Sydney magazine, some Chinese were murdered for refusing to tell where their gold was hidden, and others were buried alive as they were caught down shafts trying to hide their possessions. A white woman, who was married to a Chinese man, and her children, one of whom was a baby, were viciously attacked and would have been murdered if they had not been rescued by some of the less violent rioters. As it was, her tent was destroyed and the baby's cradle burnt. When they had tired of the violence, the rioters took to looting and destroying the Chinese tents.

When, two weeks later, the police arrested three men and charged them for their part in the riot, 3000 angry miners attacked the police camp. Four policemen were wounded and one of the attackers was shot dead. The mob dispersed when mounted policemen charged them with swords, drawing blood from many of the fleeing miners. After this incident, the New South Wales government passed a law restricting the entry of Chinese into the colony. Later in the decade and in the 1870s, thousands of Chinese travelled north to the newly opened Queensland diggings.

Most of the Chinese who came to the Australian colonies eventually returned home. Some, however, including those who could not afford to pay their passage home, remained in Australia. They worked at a variety of occupations, many of them as farm hands or shearers. Some who had done well at the diggings settled in Melbourne and Sydney, where they opened up businesses such as restaurants in what was later to become the 'Chinatown' areas of those cities.

THE EUREKA STOCKADE

Serious as the clash at Lambing Flat was, it did not compare with the terrible violence and bloodshed that was unleashed on the Ballarat goldfield six-and-a-half years earlier on the morning of Sunday 3 December 1854. In what was one of the bloodiest, and shortest, battles ever fought on Australian soil—it was over in about 15 minutes—30 miners and five policemen were gunned to death and many more were wounded. The basic cause of the conflict was the miners' resentment at the licence fees they were forced to pay and the bullying manner in which the goldfields police collected them.

Resentment at Ballarat had been building up almost from the time that miners returned to the area, after abandoning it at the end of 1851. As we noted earlier (page 19) the alluvial gold at Ballarat had quickly disappeared and miners had to dig deep for their gold. Some dug deep and struck it rich; others dug deep and ended up poorer. In any case, it was often many months before a miner, or a syndicate of miners, knew whether the would be successful. In that time they may have burrowed down 30 or 40 metres and had to work day in and day out in a deep, dark and narrow hole—each claim was less than three metres square. Often the shaft filled with water from underground streams and the miners had to bale it out with buckets winched down from above or pump it out mechanically.

Under such conditions, and with such uncertainty, it is not surprising that they resented the officials who regularly came around to check on licences and demand the monthly payment of 30 shillings. It is no wonder that many of them hid to avoid the money collectors or refused to pay, even though this could land them in gaol. One of the ruses to which they resorted was having two or more miners share a licence and pass it from one to the other when licence inspections happened.

Sir Charles Hotham.

Peter Lalor.

A new governor

In June 1854, Sir Charles Hotham replaced George LaTrobe as governor of Victoria. Hotham had been a rear admiral in the British navy and was a strong disciplinarian. When he visited the Ballarat diggings just two months after his arrival in the colony, he was given a rousing welcome by the miners. They hoped that the new governor would abolish, or at least lower, the licence fees. They were to be bitterly disappointed. Hotham soon realised that the amount of licence money being collected was too small, given the number of miners that were in the fields, so he immediately ordered the police to carry out more frequent licence checks. This, of course, made the miners at Ballarat and everywhere else in the colony even more angry.

A late night murder

Strangely enough the beginnings of the serious troubles at Ballarat had nothing to do with licences. Late on the night of 6 October 1854, a drunken Scottish miner, James Scobie, and his equally drunken mate, Peter Martin, staggered into the Eureka Hotel, the largest hotel on the Ballarat diggings, and got into an argument with James Bentley, the publican, who was an ex-convict from Van Diemen's Land. When Bentley's wife intervened in the scene, Scobie abused her with foul language. Scobie and his mate left but were pursued down the road by the Bentleys and several other men. Bentley was armed with a shovel. The pursuers set upon Scobie, but Martin escaped and made it back to his camp. He roused another man, and headed towards the

Three miners hide as police inspect licences.

The battle of the Eureka Stockade.

hotel. On the way they found Scobie dead on the ground.

Bentley, who was unpopular with the miners and rumoured to be in league with the local gold commissioner, was arrested and tried for murder. When he was acquitted, the miners assumed that he had bribed the magistrate. On 17 October a large crowd of miners rampaged into the Eureka Hotel, ransacking it and then burning it to the ground.

At about this time, an Irish digger, Peter Lalor, who was working the claim next to Scobie's, began organising meetings of miners. Lalor was a qualified engineer, who had come to Victoria in 1852. He was a commanding figure and an effective public speaker, a natural leader in the miners' cause.

Hotham's response to the hotel riot was to send 450 extra police to Ballarat and to arrest three of the supposed ringleaders and bring them to Melbourne for trial. He also ordered Bentley's rearrest and put him once again on trial. On 11 November a meeting of miners formed the Ballarat Reform League. Its aim was to stand up for miners' rights. Among other demands, it wanted miners to have the right to vote for members of parliament and it called for the abolition of mining licences. Apart from Lalor, its leaders included a German miner, Frederick Vern, and an Italian, Raffaello Carboni.

When on 20 November the three miners arrested for the hotel burning were convicted and sent to prison, the miners sent three of their number to Melbourne to demand their release. Hotham listened to the three miners' representatives courteously, promising to consider what they said, but refusing to release their gaoled colleagues. At the same time he ordered extra troops to Ballarat. When the first of these entered the town late in the day on 28 November, crowds lined the road and pelted them with stones. Even worse, a group of Irish miners rushed one of the carts carrying the soldiers and briefly took a number of them prisoner. After a violent scuffle, the soldiers escaped. Several of them were wounded and one of them later died.

Under the Southern Cross

The next day, 29 November, a huge public meeting was held at a place called Bakery Hill. Among the flags fluttering in the breeze was a ew onewith a white cross on a blue background. In the cross were five stars, representing the Southern Cross. It was the standard of the Ballarat Reform League and was a symbol of the miners' defiance of the authorities.

One of the miners' delegates who had seen Hotham in Melbourne gave a report on their meeting with the governor, highlighting his refusal to free the three gaoled miners. Peter Lalor made a rousing speech. But what really excited the crowd was a motion put by Frederick Vern, that the miners burn their licences and protect each other from arrest. A bonfire was lit on the spot and about 500 miners threw their licences into it. This amounted to open rebellion.

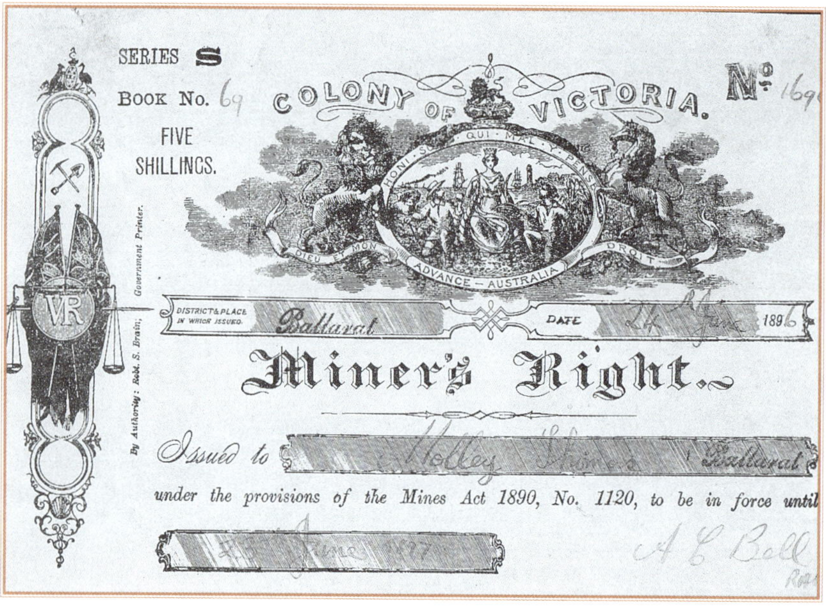
SERIES S
BOOK No. 69
COLONY OF VICTORIA.
No. 169
FIVE SHILLINGS.
By Authority: Robt. S. Brain, Government Printer.
DISTRICT & PLACE IN WHICH ISSUED. Ballarat
DATE 24 June 1896
Miner's Right.
Issued to ... Ballarat
under the provisions of the Mines Act 1890, No. 1120, to be in force until
June 1897
A. E. Bell

The 'Miner's Right', which replaced the mining licence in 1855.

The local gold commissioner, perhaps on Hotham's orders, retaliated by ordering a licence hunt the very next day. On this hot blustery day, fights soon broke out between soldiers and licence inspectors. Stones were thrown and soldiers were brought in to restore peace. Several miners were arrested. A defiant crowd left their diggings and went to Bakery Hill. There, after some argument, Peter Lalor was elected as leader of the Ballarat Reform League. Then, in a dramatic gesture, Lalor knelt beneath the Southern Cross flag and asked the miners to swear a solemn oath: 'We swear by the Southern Cross to stand truly by each other, and fight to defend our rights and liberties'. About 500 miners took this oath.

Under Lalor's command the miners constructed a rough enclosure—now known as the Eureka Stockade—from planks of wood they used to strengthen their shafts. They gathered as many weapons as they could find, went through some simple military drills and waited for Hotham's men to attack. By some accounts there were more than 1000 miners inside the stockade on Friday 1 December, but as Friday and Saturday passed without incident, many went back to their claims. Many too went back to their tents to sleep at night, returning to the stockade during the day. On the night of Saturday 2 December about 150 miners, many of them drunk on whisky, settled down to an uneasy sleep.

In the early hours of Sunday morning, about 300 soldiers and police, some of them on horseback, quietly approached the stockade. Just before dawn the soldiers stormed the stockade, firing as they came. The miners, many of whom were woken by the attack, fought back as best they could, but were soon overwhelmed, many of them before they could lay their hands on their guns. Among the dead on the government side was one officer, Captain H.C. Wise. Among the wounded on the rebels'

The site of the burnt-down Eureka Hotel.

side was Peter Lalor who received a bullet in the left shoulder.

The wounded Lalor lay hidden beneath fallen planks of the stockade as the battle raged around him. He was rescued by friends and concealed, first in the presbytery of the local Catholic church. Later, after his left arm had been amputated, he was taken to Geelong, hidden in a dray beneath a load of wood. Though a reward of £200 was offered for information leading to his arrest, he remained in hiding for several months. Vern, too, escaped capture and was never arrested.

Raffaello Carboni was among the 120 or so miners who were roughly rounded up after the disturbance, and he was one of the 13 who were sent to Melbourne in February 1855 to stand trial for treason. At these trials, however, it was argued that the government troops had deliberately provoked the miners and had behaved with excessive brutality. All but one of the accused were acquitted and set free. Only one man went to prison for his role at Eureka, and he was not a miner. He was Henry Seekamp, the editor of the *Ballarat Times*. He served six months for publishing articles that were sympathetic to the miners' cause and that were considered to be 'seditious'.

Hotham eventually had to admit defeat. He declared an amnesty for all the miners at Eureka and Peter Lalor was able to come out of hiding.

Some months later, largely as a result of the Eureka uprising, the hated mining licences were abolished in Victoria and were replaced with a 'Miner's Right', which cost a miner only £1 per year. A new tax was levied on all gold that was exported from the colony. Miners were also granted the right to vote in parliamentary elections. In November 1855, Peter Lalor was elected to the upper house of the Victorian parliament. Except for four years, from 1871 to 1875, he remained in parliament until 1887, two years before his death.

Hotham was one of the greatest casualties of the Eureka uprising. This proud man was broken by the events and by his failure to impose his will on the miners. His health declined and he died in December 1855.

THE GOLDEN WEST

Gold rushes came late to Western Australia, but when they did, they came just in time to bring prosperity to the colony and to save it from the effects of the great economic depression of the 1890s which brought misery and poverty to the eastern colonies.

If gold was not found earlier in the west it was not from want of trying. In the early 1850s, when the rushes were on in New South Wales and Victoria, Western Australia was still in the grip of an economic slump. Some concerned citizens of Perth offered a reward to anyone who could find payable gold in the area around Perth. No-one found gold and the reward was not claimed. In 1862, the government of the colony invited Edward Hargraves to prospect for gold in the west. It also offered a reward of £5000 to any discoverer of payable gold. Neither Hargraves nor anyone else was successful.

It was not until 1885 that the first payable gold was found. And it was nowhere near Perth. Encouraged by a report published by a geologist, E.T. Hardman, who claimed that there was gold in the remote, rugged Kimberley region in the far north-west of the colony, Charles Hall led a team of seven from the tiny port of Derby out into the wilderness. In 1885, at the place now known as Halls Creek, they struck payable gold.

The gold rush that followed was a disaster for many of those who joined it. Some trekked overland from the tiny settlement of Wyndham, 300 kilometres to the north; most came eastwards from the newly established port of Derby on the west coast. Hundreds even made the long and difficult journey across country from the north Queensland diggings. Many who came knew nothing of the countryside and had no bushcraft. Some died of disease, hunger or thirst before ever arriving, or on the equally difficult journey back. Estimates put the number of people at the field in September 1886 as 2000. But a couple of months later, as the wet season set in, conditions became intolerable. Disease was

The port of Wyndham, the northern gateway to the Kimberley diggings.

A gold escort in Coolgardie, 1894.

everywhere and the diggings were soon abandoned. Very few more than covered their expenses, and most lost out badly.

Rushes further south

In 1887 a 15-year-old boy on a sheep station found gold in a piece of rock he picked up to throw at a crow. The sheep station was located south of the earlier find, near the coast about 110 kilometres from the town of Roebourne. Before long the sheep station was overrun with gold-seekers, but the richest fields in the area were opened up a little later when Ben Christie found gold on the banks of Pilbara Creek. He worked the area alone for some weeks before word got out and people flocked to the area, moving further into the interior and opening up a series of fields which for some time yielded good rewards. The Pilbara goldfield, which was proclaimed in 1888, gave its name to the whole surrounding region. Fields grew up around the Ashburton River and Marble Bar and a number of rich finds were made.

Gold, however, was only one of many minerals to be mined in the Pilbara, which proved to be extremely rich in iron ore as well as having large deposits of copper and tin.

In the middle of 1890, J.F. Connelly, a Victorian who had been to the Kimberley and Pilbara fields, found gold on another sheep station. This one was further south again, about 400 kilometres north-east of the coastal town of Geraldton, at a place called Nannine south of the Murchison River. Connelly thought that the gold was not rich enough and he went elsewhere to look. But he reported his find at a remote police station in the area and soon after, word of the discovery got back to Geraldton. There was no immediate rush, but a few made their way to the Murchison region and found good alluvial gold. Connelly eventually made his way back to the scene of his original find and staked his claim there. When, some time later, one of the Murchison prospectors arrived back in Geraldton with lots of gold to show, the news spread like wildfire.

Connelly had discovered the richest field yet in Western Australia. It also had the advantage of being much easier to get to than the more northerly ones, and, being south of the tropics, of having a much milder climate. As

Western Australian goldfields.

well, kangaroos and other animals abounded in the area, and food was plentiful. Connelly worked a large claim and left the Murchison goldfields, at the age of barely 30, a rich man.

The last of the western goldfields to be opened up before the great rushes of the 1890s as found in desert—largely uninhabited and uninviting country 400 kilometres north-east of Perth. Gold was first discovered in the region in 1887, but the area did not look promising enough to develop. In 1888, however, three white men and an Aboriginal guide struck payable gold on a hill above the present town of Southern Cross, so named because the prospectors used the stars as their compass. There was no gold rush to Southern Cross—it was in the middle of waterless wastes and useless to miners without advanced equipment—but mining companies hastily set up in Perth soon had hundreds of miners working underground there extracting gold from the quartz reefs.

Riches undreamt of

Southern Cross never amounted to much, but it led the way to the richest goldfields ever found on the Australian continent. In the middle of 1892, Arthur Bayley and William Ford set out from Southern Cross, moving eastwards through arid country, looking for gold. Bayley was an experienced prospector and had won gold on the Murchison. He could afford to equip his expedition properly for the arduous journey through dry country. After travelling 160 kilometres, when their water had almost run out and they were about to return, they struck alluvial gold. When they returned to Southern Cross and left again almost immediately, other miners, sensing that they had struck gold, followed them and found their secret. Within weeks, the rush to Coolgardie, as the place became known, was on. Within months a rough frontier town had grown up around the diggings. Coolgardie became a prosperous mining centre, but for many it was merely a springboard for further explorations. It was not long before another, and greater, find was made.

In June 1893, Paddy Hannan, a middle-aged Irishman, who in his younger days had prospected at Ballarat and on South Australian fields, as well as in New Zealand, left Coolgardie with two Irish mates, Thomas Flanagan and Dan Shea. A mere 40 kilometres to the north-west, they found Australia's greatest goldfield. Hannan rushed back to Coolgardie to claim the 20 acres (8 hectares) of land that the Irishmen were entitled to by law for finding payable gold. Hundreds followed him back to the place that became known first as Hannan's Find, and then later as Kalgoorlie.

Two years later, Kalgoorlie was a booming town of 6000 people and six streets. Soon after the turn of the century, Kalgoorlie's population had risen to 30 000.

The post and telegraph office at Kalgoorlie, 1894.

An area south of Kalgoorlie, known as the Golden Mile, proved to be the richest source of gold. The township of Boulder, which is now part of Kalgoorlie, grew up around this area. It was there that investors and mining companies made their fortunes from the richest veins of quartz on the continent. Now, a century after it was discovered, gold is still being mined there, and along with nickel it is the main source of the town's present prosperity.

Gold prospectors with camels, near Kalgoorlie, 1894.

Dry work

Many an unlucky or imprudent prospector perished from thirst or hunger in the deserts of Western Australia. And for the early inhabitants of Kalgoorlie, water, and often food, were scarcer than gold. Until a railway link to Perth was opened in 1896, water and food were brought in overland on horse- or camel-drawn carts.It was often scarce, frequeny contaminated, and almost always expensive. Another source of water were the large, cylinder-shaped condensers which converted salt water drawn from underground salt lakes into fresh water. The water problem was finally solved in 1903, when a pipeline from Mundaring, 600 kilometres long, provided Kalgoorlie with a permanent water supply.

Disease was a constant danger on these goldfields. Bad food and water, especially in the early days, were a source of dysentery and there were frequent outbursts of typhoid fever, a killer disease.

To prospect for alluvial gold in this waterless environment, miners used a technique known as dry-blowing. Instead of washing out the gold from the gravel, a miner would pour soil and gravel from one dish, held up high in the air, to another on the ground. The wind would blow the dust and small particles away as they fell, allowing any gold to fall with the gravel into the lower dish. He then shook the dish to separate out any gold pieces from the dirt and gravel. Various machines, with bellows in them, were developed to make this process easier.

A population explosion

As it had in Victoria more than 30 years earlier, gold brought both people and prosperity to the colony of Western Australia. Between 1886 and 1896 its population more than trebled, from about 40 000 to almost 140 000. None of the newcomers were Chinese—the government passed a law barring them from the goldfields.

The gold rushes had another important consequence. A referendum was held in 1900 to determine whether Western Australia would become a state of the independent nation of Australia. Many Western Australians wanted the colony to remain separate, but when the vote was taken, the large numbers of miners who had come from the eastern colonies ensured that the 'yes' case would win. As a result Western Australia was part of the new Commonwealth of Australia that came into being on 1 January 1901.

A water condenser being taken to Coolgardie.

Aspects of Goldfields Life

Women and children at the diggings

Men always greatly outnumbered women on the goldfields, but even from the first gold rushes, there were women who accompanied their husbands to the diggings. Others followed once their husbands had found some gold and were able to provide something approaching reasonable living conditions. A few, but not many, adventurous women took themselves off to the diggings as independent prospectors.

The female population of the Victorian fields, in particular, increased as many miners from overseas arrived, bringing their families with them. Although they did not work the shafts and underground mines, many women at the diggings were actively involved in alluvial mining, usually helping their husbands. A lot of them, however, were deterred from doing this by the law, published in the Victorian gold regulations in 1854, stipulating that a woman, even if she were working with her husband on his claim, was required to purchase a separate licence.

The lot of most goldfields women was to be housekeepers (or often tentkeepers), washerwomen and cooks. In those days these were roles that women generally accepted, but on the goldfields they often did not. The experience of living in cold, muddy tents or crude timber huts, which were impossible to keep clean or dry, drove many a woman to abandon her husband—or to take up with another man whose winnings were greater and who could offer her a better life. Goldfields life often led to bitter domestic arguments and spelt the end of many a marriage.

Among the women who came on their own to the goldfields were a number who set up in business—as storekeepers, or as laundresses who did washing for miners. Women—some of them miners' wives—who provided miners with illegal alcohol by running sly grog tents, were disapproved of by the authorities, and by many other women. But they were well patronised, and often protected from the police by the miners. So too were those who served as prostitutes to lonely men who would willingly pay for their services. As pubs grew up around the New South Wales fields, young women came to work in them as barmaids. Others worked as dancers in the dance halls around the diggings, offering entertainment and companionship to the diggers in exchange for money or gold.

A visitor to Kalgoorlie in 1895 observed that there were few women at the Western Australian diggings. But there were women there, many of them from South Australia, working behind the scenes as nurses at the makeshift hospitals at Coolgardie and Kalgoorlie. They were caring as best they could for the many victims of dysentery, typhoid and mining accidents that these frontier diggings produced.

By the end of 1852 there were 12 000 children on the Victorian diggings. Children of 10 or more were usually put to work helping to prospect for gold or assisting with cleaning

Lola Montez.

and cooking. There were some schools, often housed in large tents, which could hold up to 100 students. As towns grew, so did more permanent schools. But education for most goldfield children was haphazard at best. Teachers, usually poorly trained, came and went and there was very little equipment. Even though parents paid for their children to receive lessons, education was often very disjointed, as families moved from one goldfield to the next and children were forced to change schools. One inspector of schools in Victoria wrote despairingly of an 11-year-old girl who had already been to six schools and who had so far learned nothing at all. She was probably fairly typical.

Children cradling for gold.

The state of children's education on the goldfields was well summed up by the poet and balladeer Charles Thatcher, who wrote:

> Here the boys grow up without
> The slightest education
> To teach them to succeed in life,
> In following their vocation:
> Of course they cannot read at all,
> Nor can they write their names,
> And the only mark they ever make
> Is when they 'mark' our claims.

Entertainment

A digger in Bendigo in the 1850s wrote that 'amusements are not in harmony with the diggings. Men come there usually to work in earnest, and they have no time for play'. But then he added 'yet now and then a song is heard, with the notes of a flute or violin ... In one tent near me there was an occasional concert of a fife, a dish-bottom drum, and a primitive sort of triangle ...' Life on the goldfields was mostly work, but there was also a fair amount of play. For much of the time, miners had to make their own entertainment. They found time to relax on Sundays, and those who had money or gold could repair to the pubs or sly grog tents on Saturday, and other, evenings. Dance rooms, often attached to hotels, were popular meeting places, even though the men often had to dance with each other, to the accompaniment of a scratch band, when women were in short supply. Improvised concerts and boxing matches were also popular forms of impromptu entertainment.

As gold towns were established, however, theatres and other forms of professional entertainment became more common. Probably the first organised entertainment on any goldfield was provided on the New South Wales Turon goldfields in 1851. When Henry Burton, whose circus had been performing in Maitland, in New South Wales' Hunter Valley, heard of the Turon finds, he promptly packed up and drove his horses and carts over the mountains. For six months his circus performed to the miners at the diggings there. Then, as glowing reports of the Victorian diggings came to hand, Burton packed up again and headed south to Victoria, where he hoped to attract even larger audiences and greater takings.

As gold towns developed, so did the range of amusements they offered to diggers who could afford to enjoy them. Horse races became a regular form of entertainment around many goldfields. Betting on them gave many a gold-digger an opportunity to wager, for better or worse, the winnings he had worked so hard to get at the diggings. Theatres grew up, offering quality productions, often with noted performers from overseas, of well-known plays as well as dancing and other forms of variety entertainment. In 1854 in Bendigo there were already three theatres, all attracting good audiences.

Probably the most famous entertainer to perform to Victorian diggers was Lola Montez, who, despite her assumed Spanish name, was really Irish. She achieved a great success at Ballarat early in 1856 with her 'spider dance'. In what she claimed, not very convincingly, was a 'traditional' Spanish dance, she whirled around wildly, pretending to search for a spider that was hidden in her skirt. The diggers loved it; many people thought it highly immoral. But the highlight of her visit was a disagreement she had with Henry Seekamp, the editor of the *Ballarat Times*, who had earlier gone to gaol for his support of the miners at the Eureka Stockade. When Seekamp attacked her in his paper, not for her performance but for details of her private life, Montez sought him out in a Ballarat hotel and took to him with a whip. Seekamp, who had a riding whip with him, retaliated and the two combatants had to be pulled apart before they did serious injury to each other.

A boxing saloon at Ballarat provided entertainment for diggers.

INDEX

Aborigines, 36, 38
Agamemnon, 7
'Alexander the Yank', 29
Ancient Egyptians, 7
Anti-Chinese feelings, 47-50
Appis Creek, 32
Bakery Hill, 54
Ballarat, 17-19, 21, 24, 43, 51
Ballarat Reform League, 53-54
Bathurst, 7, 8, 12, 14
Bayley, Arthur, 58
Beechworth, 24-25
Bendigo, 21, 22-24
Bentley, James, 52-53
Beyers, Louis, 31
Burton, Henry, 62
Bushrangers, 44-45
California, 10-11
Campbell, 16, 17
Canvas Town, 41
Carboni, Raffaello, 53, 55
Carter, Howard, 7
Castlemaine, 21
Cecil, Lord Robert, 42-43
Central Deborah mine, 22
Chapman, William, 37
Charters Towers, 32, 34
Children, 61
Chinese immigrants, 38
Christie, Ben, 57
Clacy, Ellen, 41-43
Clarke, William, 8, 9
Clunes, 17
Connelly, J.F., 57
Cooktown, 36-37
Coolgardie, 57-58
Day Dawn mine, 32, 34
Dickens, Charles, 38-39
Dunlop, John, 18
Eaglehawk Gully, 24
Early finds, 8-9
Echunga, 37
El Dorado, 7
Entertainment, 62
Esmond, James, 17
Eureka Hotel, 52-53
Eureka Stockade, 51-55, 63
Fauchery, Antoine, 26-28, 42, 47
FitzRoy, Charles, 12
Flanagan, Thomas, 58
Forbes, 29, 32
Ford, William, 58
Forest Creek, 21
Frencham, Henry, 23
Gardiner, Frank, 29, 32, 45
Gibson's Plains, 29
Gipps, George, 9
Gold fever, 7, 14-15
Gold rush, 12
Golden Mile, 59
Goldfields life, 61-63
Goldfields, troubles on, 42-45
Golden Point, 18
Gulgong, 29-30
Guyong, 8, 12
Gympie, 34
Hall, Ben, 23, 29, 45
Hall, Charles, 56
Halls Creek, 56
Hann, William, 35
Hannan, Paddy, 58
Hargraves, 14
Hargraves, Edward, 8, 10-14
Hill End, 31
Hiscock, Thomas, 17
Hodgkinson River, 36
Holtermann, Bernard Otto, 31
Hotham, Sir Charles, 51-55
Howitt, William, 42
Immigration, 38-41
Irving, Jemmy, 14-15, 20
Johnson, William, 23
Kalgoorlie, 58-60
Kennedy, Margaret, 22
Kerr, Dr William, 14-15
Kiandra, 29, 35
Kimberley goldfields, 56-57
Lalor, Peter, 51, 53-55
Lambing Flat, 29, 32, 48, 41
Lansell, George, 25-26
LaTrobe, Governor Charles, 16, 19, 23, 52
Lewis Ponds Creek, 8
Lister, John, 8, 12-14
McBrien, James, 8
McIvor goldfield, 44
Martin, Peter, 52
Melbourne, 16
Michel, Louis, 17
Miner's Right, 54-55
Moliagul, 27
Montez, Lola, 63
Mosman, Hugh, 35
Mount Alexander, 20-21, 23, 38, 42
Mount Buninyong, 17
Mount Korong, 25
Mulligan, James Venture, 36
Murchison goldfields, 57-58
Murchison, Sir William, 8
Murphy, Jeremiah, 45
Mycenae, 7
Nannine, 57
Nash, James, 32-33
Nashville, 34
New South Wales goldfields, 15, 28-31
Omeo, 24-25
Ophir, 12-14, 18, 28
Palmer River, 35-37
Peters, Christopher, 20
Photography, 26-27
Pilbara goldfield, 57
Pine Creek, 37
Pyrenees Mountains, 16, 17
Queensland goldfields, 32-37
Raikes, Bob, 43
Ravenswood, 34-35
Red Hill, 31
Regan, James, 18, 19
Sandhurst, 22
Saunders, Tom, 30
Schliemann, Heinrich, 7
Scobie, James, 52-53
Seekamp, Henry, 55, 63
Shea, Dan, 58
Sierra Nevada mountains, 10
Sofala, 14
Southern Cross, 58
Southern Cross flag, 54
Spring Creek, 24
Strzelecki, Paul Edmund de, 8
Tambaroora, 14, 31
Thompson, William 'Abednego', 22
Thomson, Deas, 8
Tom, James, 8, 12-14
Tom, William, 8, 12-13
Townsville, 34
Trollope, Anthony, 29-30
Turon, 28, 31, 62
Tutankhamen, 7
Ur, D'Arcy Wentworth, 37
Vern, Frederick, 53-54
Victoria mine, 22
Victorian goldfields, 16-17, 26
Warrandyte, 17
Wedderburn, 25
Weitemeyer, Thorvald Peter Ludwig, 37
Welcome Stranger nugget, 27
Western Australian goldfields, 56-60
Wise, Captain H.C., 54
Women and children, 61
Wright, William Henry, 21
Young, 29, 32
Young, Sir John, 29